AF306383

Force Estimation and Control in Human-Robot Interaction

Topics in Systems Engineering

Print ISSN: 2810-9090
Online ISSN: 2810-9104

Series Editors: Luigi Fortuna *(Università degli Studi di Catania, Italy)*
Arturo Buscarino *(Università degli Studi di Catania, Italy)*

The Series aims to cover a wide spectrum of Engineering topics but with a strong characteristic of interdisciplinarity using less technical items. It could be a series including thinking aspects, history topics linked with engineering topics from civil engineering to industrial engineering and so on. The contributions will cover a wide series of engineering topics and could be useful both for educational purposes and for research information. The idea is to combine technology, science, arts and social science with some emerging subjects.

Topics in Systems Engineering — Volume 5

Force Estimation and Control in Human-Robot Interaction

Adriano Scibilia

STIIMA-CNR, Italy

World Scientific

NEW JERSEY · LONDON · SINGAPORE · BEIJING · SHANGHAI · HONG KONG · TAIPEI · CHENNAI · TOKYO

Published by

World Scientific Publishing Co. Pte. Ltd.

5 Toh Tuck Link, Singapore 596224

USA office: 27 Warren Street, Suite 401-402, Hackensack, NJ 07601

UK office: 57 Shelton Street, Covent Garden, London WC2H 9HE

British Library Cataloguing-in-Publication Data
A catalogue record for this book is available from the British Library.

Topics in Systems Engineering — Vol. 5
FORCE ESTIMATION AND CONTROL IN HUMAN-ROBOT INTERACTION

ISBN 978-981-98-1361-2 (hardcover)
ISBN 978-981-98-1362-9 (ebook for institutions)
ISBN 978-981-98-1363-6 (ebook for individuals)

For any available supplementary material, please visit
https://www.worldscientific.com/worldscibooks/10.1142/14327#t=suppl

Desk Editor: Muhammad Ihsan Putra

Typeset by Stallion Press
Email: enquiries@stallionpress.com

Preface

The field of human–robot interaction (HRI) represents a convergence of engineering, mathematics, and cognitive sciences, where the limits of what machines can achieve in collaboration with humans are constantly being pushed. This book seeks to explore the dynamic relationship between humans and robots, focusing on modeling and control strategies that can effectively characterize, simulate, and predict human control behavior while interacting with a controlled machine, enabling seamless and effective collaboration in shared environments.

Central to this exploration is the idea that a successful HRI requires both a deep understanding of human behavior and the possibility to design responsive robotic systems that are able to use such information in the most effective way possible. This balance is achieved through the interplay of structured mathematical models and modern data-driven techniques. By combining these approaches, it becomes possible to predict and adapt to human actions in real time even in uncertain and dynamic scenarios.

The journey begins with an introduction to the human modeling efforts provided in the scientific literature, starting from the aeronautic domain, where the interest in this topic has been raised in the first decades of the last century to understand and predict pilot motions when controlling an aircraft. These foundations provide a basis for understanding the complexities of collaborative tasks, where physical interaction demands both precision and adaptability. As we will see, the findings obtained in this research area, which apparently seem so far from modern HRI, can be successfully applied in robotics to obtain a first raw understanding of human control behavior.

As the focus shifts to nonlinear models, the challenge of estimating human intention and forecasting trajectories is addressed by considering inevitably more complex system dynamics from both the perspective of the controlled machine and the human subject that is interacting with it and, therefore, adapting to it. Here, the integration of machine learning techniques, such as artificial neural networks, with classical modeling frameworks is shown to play a crucial role in bridging the gap between theoretical understanding and practical implementation in many technological contexts apart from robotics, such as autonomous driving or unmanned aerial vehicles.

Going another step forward, the book shifts its focus from the general human–machine context of our multi-disciplinary analysis to a specific area in which we will land: human motion prediction in robotics. The evolution of the developed methodologies did not follow the path of increasing the complexity level of the represented dynamics, moving from model-based to data-driven algorithms, as done previously, but rather finding a compromise between them. The resulting hybrid approaches, combining both of these philosophies, proved to be the right choice to capture all the aspects involved in a complex scenario like the one we are studying.

This book is based on the work done during my doctoral studies and is inspired by my PhD thesis. It's intended as both an academic reference and a practical guide for researchers and students in robotics, control systems, and artificial intelligence. Our goal is to offer a combination of foundational theory, methodological innovation, and real-world applications to provide the reader with the insights necessary to implement and put into practice state-of-the-art methodologies in a robotic control system. The author invites readers to engage with the challenges and opportunities that lie ahead in this rapidly evolving discipline.

Adriano Scibilia

Contents

7. Frequency Domain Analysis and Human Reactive Delay Estimation   117

8. Peak-to-Peak Dynamics   127

Chapter 1

Introduction

In any system characterized by a close human–machine physical interaction, providing controlled elements with the ability to identify and understand what the human operator is doing is crucial to increase efficacy and safety. While this is an ability that humans naturally learn over time, machines need to be explicitly trained on how to do this. Such recognition problems are heightened by the dissimilarities between humans and controlled machines from a mental, computational, and physical point of view. These differences imply that, when faced with the uncertainty of the real world, machines cannot always count on humans to behave as expected and cannot always easily anticipate how they will react to an unexpected event (Hiatt *et al.*, 2017). One way this challenge can be addressed is by equipping machines with explicit models of their human teammates. Many different techniques are used to model human cognition and behavior, spanning different timescales and levels.

Modeling of the human control action when interacting with a controlled machine has become almost an independent research field over the years, involving multiple disciplines and approaches. Neurophysiologists and cognitive scientists have considerably improved their understanding of human perception, information processing, and control strategies with respect to prior approaches, mainly focused on qualitative descriptions of possible human decisions and actions. Since the instrumentation and measurement techniques have dramatically improved, as well as the power of computational calculations, scientists have developed functional maps of neurons and identified deep brain functions (Bergman, 2016; He *et al.*, 2020; Markram, 2012).

Still, the human brain's dexterity and plasticity have a lot of mysteries, and how humans can interact and adapt themselves to unknown external dynamics is an open issue. Therefore, intense research was put into investigating motor control internal dynamics, perception, and learning. With particular reference to the study of input–output characteristics of the motor apparatus, the concept of the internal model allowed significant advances for describing human adaptation to external dynamics and trajectory planning (Kawato *et al.*, 1987). The concept underlying the internal model hypothesis found its origin in robot control. In fact, a robot needs knowledge about its internal kinematic model in order to perform any position or velocity control. The same concept was extended to human physiology when Ito (2000) proposed that internal models of the limbs and connected brain regions are present in the cerebellum. The acquisition of inverse dynamics of the motor efferent systems, along with the inverse dynamics of the controlled object, helped us to explain how it is possible to perform fast and complex movements even if time delays and low gains characterize biological feedback structures (Jordan and Rumelhart, 2013). An opposite approach for the same problem relies on equilibrium-point control models (Feldman and Levin, 2009), in which the central nervous system is able to control muscle dynamics by simply acting on its threshold level.

Predictive simulations of human movement were also used to dissociate the contributions of neural and musculoskeletal impairments to gait deficits in cerebral palsy (Falisse *et al.*, 2020), evidencing the importance of a precise model of involved body parts and physiological districts to improve the correspondence between simulated and measured data (De Groote and Falisse, 2022). The same was found in the rehabilitation robotics field, where mechanical impedance control parameters of the upper limbs of a human were identified in order to adapt the rehabilitation robot's training strategy (Demir *et al.*, 2016; Zou *et al.*, 2022) accordingly. Modeling approaches of this type linked mathematical descriptions typical of classical control theory with a functional description of physiological systems acting during the control process. This was able to link neurosciences with more practical engineering fields such as robotics and aerospace. Aerospace researchers had the first application scenario, which motivated an interest in this topic in the first decades of the last century. Starting from World War II, engineers' and psychologists' efforts were directed toward the modeling of human behavior as an inanimate feedback controller to improve the performance of pilots and bombardiers (Tustin, 1947). The first approach was to consider the human controller as an inanimate servomechanism, which can be represented in a simple feedback structure as a block with well-defined input(s) and output(s). In a manual human-in-the-loop control problem, the input of the human is an error signal, usually visual. Its output is human control actions, which provide a command to the controlled element (i.e., a gun, an aircraft, a vehicle, etc.). The same consideration is still valid in physical human–robot interaction, especially in those applications involving force or impedance control strategies. Here, continuous and compliant physical contact is required in order to perform cooperative kinds of tasks, which can be reduced to mutual movement between humans and robots. In this case, the human operator should be able to deviate from the robot's initial trajectory and impose a different one.

In such a context, the human operator will act as a proper motor controller by internally deciding the goal trajectory to perform and imposing an external force to achieve such an objective by manipulating the robot. From this point of view, the performance of the human operator can be well approximated as the action of an inanimate controller. This situation results in a simple compensatory manual control system. Hess (2018) explained this concept by giving the example of a human soldier performing a tracking task, attempting to keep a moving target within the gun's field of view. In this case, the angular error between the target and the azimuth of the gun's viewfinder can be considered as input, while output control action is a force acting on a simple gear mechanism. Since the soldier is modeled as an inanimate servomechanism, the mathematical representation that was used to describe him should be the same as that used to describe a linear servomechanism: a set of linear differential equations with constant coefficients, or, equivalently, as a transfer function in the frequency domain. The most famous example of this kind of approach is McRuer's crossover model, in which the human is represented as a general quasi-linear describing function. From this first approach, the development of mathematical descriptions of human controllers with a control-theory approach evolved along with new control techniques. For instance, the development of linear quadratic Gaussian control systems (LQR), when applied to human operator's modeling, led to the "optimal control model." The same concept is valid for recent modeling techniques, such as fuzzy control models or models based on neural networks, which followed the spread of these techniques in systems design.

However, human modeling efforts, which started from this applicative scenario, have successively proved to be useful in many other domains of the engineering field. Classic examples are the design of display and control equipment interfaces based on man-machine environment system engineering in various types of plants (Wang *et al.*, 2021a; Wu *et al.*, 2020) and, more recently, service robotics applied to healthcare (Ngo *et al.*, 2020, 2022). The understanding and prevision of human action have been extensively investigated in human–robot interaction (HRI) in the last few years, being considered as input information to be gathered that enables compliant and adaptive behavior of the robot. Khoramshahi and Billard (2019) developed a task-adaptation framework to allow robot compliance with respect to human movements. Similarly, Neto *et al.* (2019) discussed gesture-based HRI, allowing for an automatic task manager parametrization where the human can help or correct a robot's choices in collaborative assembly applications. Furthermore, Desideri *et al.* (2019); Ullrich *et al.* (2021) tried to analyze the psychological and emotional implications of continuous interaction with a robot in a production setup. Behavioral criteria were also considered in HRI in commercial vehicles (Vasconez *et al.*, 2019), where a scheme of mental state variables can be used to modulate driving velocity and braking in different moments of the day and night. In a similar context, machine learning techniques can also be applied to HRI intelligent transport systems (Wang *et al.*, 2022a).

With most of these models being the result of highly application-specific efforts, their variability causes some difficulties when trying to find common features and divide them into general categories. In other words, every effort put into the definition of human control models originated from the need to describe their behavior in a particular situation. This was done starting from different perspectives and with different levels of abstraction. According to Rasmussen (1983), human behavioral models, when interacting with an aircraft, can be grouped into three types: skill-based, rule-based, and knowledge-based. In skill-based models, the human–machine system is continuously controlled following a mission statement. Rule-based models provide a discrete decision-making description of human behavior, which is guided by a stored rule. The last category groups all the control strategies deriving from unexpected events or unfamiliar environments in which the human operator has to avoid dangers and risks. An example can be found once again in the pilot's control of an aircraft. Traditional control models are mainly part of the first category, while the development of modern artificial intelligence techniques has increased the use of models based on the other two. Xu *et al.* (2017) proposed a different classification, identifying human models based on control theory, based on human physiology and intelligence techniques. Classical feedback control models, such as McRuer's crossover model and the optimal control model, fall into the first category.

Successive modeling techniques in which a better description of all the underlying processes that determine the overall human control strategies were Hess's structural model and Hosman's descriptive model. Hess proposed a detailed description of human perception processes and inner feedback loop, while Hosman modeled the interaction between visual and vestibular inputs and their influence on the overall control strategy. Additional models that can be associated with this category are biodynamic models, which try to include biomechanical effects of the body moving into an accelerating environment such as an aircraft or a vehicle. Equivalent to the previously described knowledge-based models, the models based on intelligent techniques include approaches dealing with uncertainty, such as fuzzy logic and neural networks.

In our analysis, the focus will be directed toward models based on control theory, including the involved physiological structures. Moreover, humans have both linear and nonlinear behavior when interacting with a machine. This aspect is reflected in the classifications of control models, where both linear and nonlinear dynamics are described.

In the first part of this work, the focus will be mostly on linear models. The second chapter will be structured as follows: In the first section, the main motor control theories will be addressed. Then, the second and third sections will detail the modeling approaches of neuromuscular dynamics and human sensory systems, respectively. The fourth and last section provides a description of the most important human–machine interaction models, in which the dynamics described in the previous sections are represented within a general control structure.

In the third chapter, starting from the state-of-the-art models described before, we propose a linear modeling strategy: the "precision model of the human–robot complex." As can be deduced from the given name, such modeling technique described the response of the whole system comprising both the human and the controlled robot. The simulated system response was compared with the measured response and with the response obtainable by simulating the system with a standard crossover model. Overall, the main application in relation to which the model was studied was the experimental identification of human reactive delay in a manual guidance task using the small collaborative UR5 robotic arm, where an external virtual reference (acting as a forcing function) was applied.

The second part of this work proceeds with the analysis by going into the details of nonlinear modeling strategies. Despite the successes, linear models lack in representing nonlinearities typical of human control behavior, especially when facing high-complexity scenarios. McRuer and Hess described the evidence of a pulsive behavior of the pilot when the demanded task is too complex, leading to the formulation of dual-loop control models. They describe human bimodal control behavior, focusing on the error compensation (typical of classical crossover theory) and visual rate sensing (used in pursuit tasks with predictable inputs) (McRuer and Krendel, 1974). The dual-channel structure proved to be more suitable for capturing nonlinear dynamics in the pilot system during the information processing stage, represented by thresholds and saturation elements, which were used for describing phenomena like pilot-induced oscillations (PIOs) (Bucolo *et al.*, 2020) or spatial disorientation (SD) (Lone and Cooke, 2014).

In such a modeling technique, the human is considered a controller, an element part of the control loop (human-in-the-loop control). Its sensing elements and muscle actuators' dynamics are related to the external stimuli, the executed task, and the controlled element. While executing a specific task, the human subject tries to optimize its behavior to achieve its goal while reducing efforts. If the difficulty increases, nonlinear dynamics is increasingly observable. Neuromuscular dynamics can be considered one of the primary physiological sources of nonlinearity in human control action. Modeling techniques in this context are often based on optimal control theory, trying to identify the system's objective function that the human tries to optimize while executing a specific motion.

Aside from classical model-based approaches, a deeper focus on information processing and learning abilities is necessary to have a complete overview of the human as a controller. Different modeling and data-driven approaches have been proposed with this goal in many research efforts, even resulting in combinations of them. The model proposed by Xu *et al.* (2021), for example, studied the origin of nonlinear PIO, proposing a multi-loop human pilot model during a multi-axis control task. Here, the pilot's ability to sense a changing situation, being based on experience and judgment, is represented by a fuzzy logic control element, able to modulate his strategy and, indirectly, the system input/output characteristics (through the variation of model parameters). Apart from multi-loop models, fuzzy logic techniques have been used in association with other nonlinear system modeling approaches and control techniques in order to deal with uncertainties in the external environment (Han *et al.*, 2023) or in model parameters tuning (Ray *et al.*, 2023). For instance, fuzzy systems and artificial neural networks (ANNs) have been successfully used in hybrid models in the past for human operator tuning (Miyoshi *et al.*, 1993) or parameter optimization of the controlled plant (Dey *et al.*, 2020). The spread of such kinds of hybrid models led to the development of neuro-fuzzy systems, which will be discussed in detail in Section 4.3.3.

Due to their simple mathematical structure and low computational cost when implemented, ANNs have been successfully used in the presence of unstructured data in learning, classification, and prediction algorithms in computer vision (Shi *et al.*, 2022), autonomous driving (Huang and Fu, 2022), bio-informatics (Zhang *et al.*, 2022c), as well as in medical (Konnaris *et al.*, 2022; Oyelade *et al.*, 2022), industrial (Rusanovsky *et al.*, 2022), and rehabilitation (Haji Hassani *et al.*, 2022; Ma *et al.*, 2022) robotics.

In human–machine interaction, however, it may be important to capture temporal relationships between raw data in order to identify the system model accurately. Special kinds of ANNs, such as the recurrent neural network (RNN) (Bartolomaeus *et al.*, 2022; D'Anniballe *et al.*, 2022) and the long short-term memory (LSTM) network (Wen *et al.*, 2022), are very common for this purpose, thanks to their internal loops between the hidden layers, which in the case of LSTM allows for capturing even long-term temporal relations. Further details on these two architectures will be given in Section 4.4, as well as other data-driven approaches, such as reinforcement learning, useful to model human decision-making and the generation of its internal goal. Classical supervised, unsupervised, and semi-supervised learning methods are introduced to represent how a "human controller" creates a strategy to achieve a long-term goal, passing through several intermediate steps. There is a wide variety of practical applications exploiting such techniques (Alabdullah and Abido, 2022; Dai *et al.*, 2022; Wu *et al.*, 2022; Zhong *et al.*, 2021). Remarkably, the optimal control theory modeled the human control action by identifying its internal cost function to minimize, similar to the reinforcement learning approach. Indeed, in such a case, the human decision-making process is modeled by describing its objective function, which is maximized by the subject during its actions.

The mentioned modeling techniques, from the classical control theory to the modern data-driven approaches, have succeeded in representing a different aspect of the human control strategy when interacting with a machine. Application scenarios such as intelligent transport systems or human–robot collaboration offered many examples of modeling and control techniques, which have been developed by combining two or more of these approaches.

This latter concept has been used to construct a nonlinear auto-regressive model of the human subject, with moving average and exogenous input (NARMAX). NARMAX models are the nonlinear evolution of the famous linear ARMAX models (Johansen and Foss, 1993) and became a gold standard on how to construct models from complex and partially unknown systems (Chen and Billings, 1989). The proposed NARMAX model was constructed by using an ANN for approximating the nonlinear functional relationship between input data, using the universal approximation property of NNs.

In Chapter 5, the human model was defined and trained in the first part using the same dataset acquired in the previous proposed linear model of the third chapter. In the second part of the chapter, the nonlinear human model was applied in a different setup using a high-payload Comau NS16 robotic arm. This way, the model performance with new and different data can be discussed and compared to the ones obtained with the first dataset. Most importantly, the model is now used online, also testing its speed and practical usability. The fact that the Comau arm is not a proper "collaborative robot," built exactly for the purpose of performing tasks with human subjects around, makes this scenario the perfect case in which the controlled plant would have a great benefit in gaining information about the human user, to estimate its intention and better collaborate with him.

In the following chapter, we will see how the proposed model can be used, due to its accuracy, to extract the delay information from data (as previously done with the precision odel) in a fast and simple way, exploiting the knowledge that we have about the controlled element's dynamics.

In the last chapter of this book, the model's ability to forecast the next peak of the output signal is analyzed. The regular presence of peaks in human control output (which corresponds to its applied force) makes it really interesting to evaluate such an aspect to study the eventual presence of peak-to-peak dynamics (PPD) (Candaten and Rinaldi, 2000). As we will see in more detail in the last chapter, PPD has been described and studied extensively in a wide range of chaotic systems and models, being able, when observed, to reduce the model's order without loss of accuracy.

Chapter 2

Linear Models

2.1 Motor Control in the Central Nervous System

Motor control dynamics in the human nervous system have been widely studied by neurophysiology researchers over the last few years with different approaches. One of the most promising assumes the existence of internal models of sensory-motor output dynamics in the central nervous system (CNS). Miall *et al.* (1993) suggested that the cerebellum forms two different types of internal models. One of them is a forward predictive model of the motor apparatus (e.g., limbs and muscles), providing a rapid prediction of the sensory consequences of each movement. The second is a model of the time delays in the control loop (due to receptor and effector delays, axons, conductances, and cognitive processing delays). This second model delays a copy of the rapid prediction so that it can be compared in the temporal register with actual sensory feedback from the movement. Both models can coexist and form two Smith predictors. Wolpert *et al.* (1995) experimentally verified that human subjects were able to estimate the hand position without visual feedback and with applied external disturbances, supporting the evidence that the CNS internally simulates the dynamic behavior of the motor system in planning, control, and learning. The existence of such internal models of motor dynamics and temporal delays in the CNS has been discussed a number of cognitive science and neurophysiology fields. The necessity for internal models in motor control has been one of the central issues of debate in relation to other approaches to motor control theory, such as the equilibrium-point control (Feldman and Levin, 2009).

In the equilibrium-point control or threshold control theory (TCT), motor actions are controlled by changing neuro-mechanical parameters, which establish the steady state (equilibrium point) and is commanded to lower levels (e.g., muscles and limbs) by descendent systems. The neural control variables that determine the equilibrium point are identified in the λ model (Feldman *et al.*, 2007). As known, muscle activation by efferent neurons and motor units is triggered by a variation in muscle length. When a muscle is quasi-statically stretched, the potential in the motoneuron membrane increases, and after that, a certain threshold value is reached, and the motoneuron starts to be recruited. Physiological data indicate that such threshold length value comprises various factors besides its central component. If the central component is λ, the composite value is

$$\lambda_c = \lambda - \mu\omega - \rho + \epsilon(t), \tag{2.1}$$

where μ is a temporal parameter related to the dynamic sensitivity of muscle spindle afferents, ω is the velocity of change in the muscle length, ρ is the shift in the threshold resulting from reflex inputs (such as those responsible for the inter-muscular interaction and cutaneous stimuli), and $\epsilon(t)$ represents temporal changes in the threshold resulting from intrinsic properties of motoneurons. In equation (2.1), the CNS controls both the λ and μ parameters. Therefore, according to the position TCT theory, high control levels can control muscular activation, minimizing the difference between the actual length and the one established by the threshold. Let's consider a situation in which the human subject is asked to hold an object; according to TCT theory, the gripping force is set in such a way that the difference between the threshold length (established by the physical shape of the object) and the actual, which is set to be virtually inside the object, is minimum. This operation results in the object being held using the minimum quantity of gripping force. The same situation, as shown in Figure 2.1, was considered from another point of view by Kawato (1999). Here, the coordination of reaching and grasping, which allows using the minimum grasping force to prevent slip when lifting an object, was considered proof of the existence of the limbs' internal inverse and forward models. When the arm grasps an object, the inverse model of the combined dynamics of the arm, hand, and object computes the necessary motor commands from the desired arm trajectory. Such commands are sent to the arm muscles and to its forward dynamic model, which can predict the future trajectory and establish the grip force necessary to lift it, considering its friction and a certain safety factor.

Actually, alternative explanations of how such predictive capabilities can be possible without internal models also exist. They are linked to biological systems' "strong predictive" and anticipatory properties. Dubois (2002)

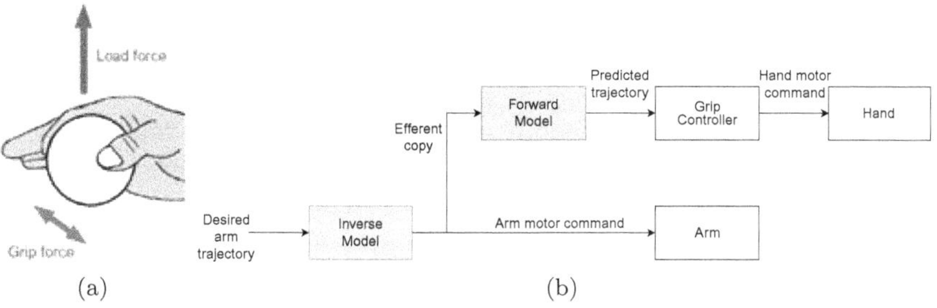

Fig. 2.1. Example of how a control structure based on internal models in the cerebellum can be used to explain human coordination of load and grip force in a simple grasping task, as studied by Kawato (1999).

defined strong predictive systems as those in which predictive properties are inherent in the systems' natural dynamics and thus do not rely on internal models. At the same time, weak predictive systems are based on internal models of themselves. Another motor-control case study that was analyzed to highlight the differences between the two approaches is the formation of an arm trajectory (Burdet *et al.*, 2006). It is known that muscles and peripheral reflexes control loops have spring-like characteristics, which can pull back the limb's joints to their equilibrium positions. This is done by generating a force directed against the sensed external perturbations. Such viscoelasticity can be considered as the static gain of the peripheral feedback control loop and can be adjusted by properly setting the associated muscle co-contraction level and the reflex gain.

The equilibrium-point control hypothesis implies that through this viscoelasticity, the brain can control the movements of the limbs by simply setting a series of stable equilibrium positions aligned along the desired trajectory (Bizzi *et al.*, 1984; Feldman, 1966). An experimental study of this concept has been presented by Flash (1987), where the data from Gomi and Kawato (1996) were reinterpreted in an equilibrium control fashion with a straight equilibrium trajectory. However, this approach requires viscoelastic forces to increase proportionally to the movement speed since the dynamic forces exerted on multi-joint links depend on the square of the velocity. Differently, the alternative explanation implying the internal model control allows the realization of accurate and fast movements even considering low viscoelastic forces (Gribble and Ostry, 1999; Katayama and Kawato, 1993; Latash and Gottlieb, 1991). Experimental evidence of a relatively low stiffness observed during movements performed by a well-trained subject has supported the latter hypothesis (Milner and Cloutier, 1993; Morasso and Schieppati, 1999). Another step forward was integrating the two approaches, muscle viscoelasticity and internal models, through computational models to learn the behavior and applications of internal models efficiently (Katayama *et al.*, 1998; Sanger, 1994). For example, Morasso and Schieppati (1999) showed that intrinsic muscle stiffness is not strong enough to stabilize upright posture; using internal models was suggested as an alternative explanation.

2.2 Neuromuscular Dynamics Model

The neuromuscular system dynamics model has been widely investigated in manipulation tasks starting from the early 1960s. Typically, muscles and manipulators are considered as a unique function. An example of what can be involved in the muscle-manipulator dynamics was studied by McRuer (1965); a simple block scheme of neuromuscular subsystems is shown in Figure 2.2. Retinal and central equalization transfer function changes according to the considered forcing function dynamics. This block was represented by simple gain and delay factors by McRuer (1965) for a rate stimulus, but changes for other controlled elements. The alpha motor neuron command α_c is the command input from higher centers down to the spinal cord. There, the change in the average firing rate of the alpha motor neurons involved is proportional to the effective driving force. The commanded force signal then goes to the muscle/manipulator block, whose dynamics, as mentioned, are represented by a unique transfer function that consists of a third-order system with one real root and a quadratic pair plus

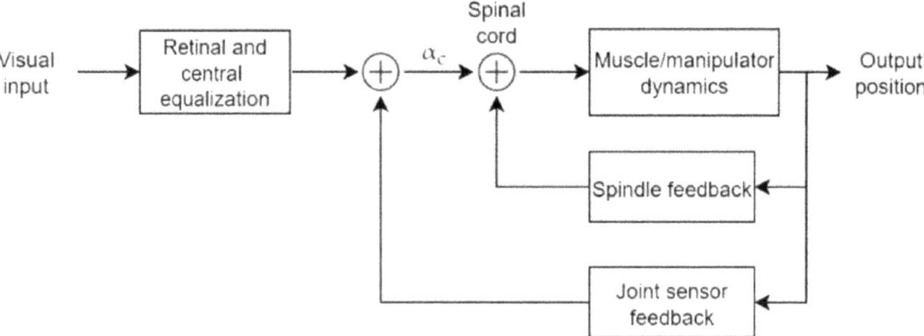

Fig. 2.2.   Model of subsystems contributing to neuromuscular dynamics in manipulative control tasks, as studied by McRuer *et al.* (1965).

a time delay, where

$$H_{MM} = \frac{-K_I e^{-\tau_\alpha s}}{(1 + \tau_N s)\left(1 + \frac{2\xi_\alpha}{\omega_\alpha}s\right) + \left(\frac{s^2}{\omega_\alpha^2}\right)}. \tag{2.2}$$

The muscle characteristics are functions of the steady-state isometric tension of the muscle system operating point. The changes in this average tension are caused primarily by changes in the gamma motor neuron system discharge. The effects of the gamma neuron bias signal, while not shown explicitly in Figure 2.1, are used to set up the spindle feedback operating point equalization, whose block also approximates the Golgi tendon force feedbacks, and the corresponding describing function is

$$H_{sp} = \frac{K_{sp}(s + Z_{sp})e^{-\tau_{sp}s}}{(s + P_{sp})}. \tag{2.3}$$

The effective joint sensor provides A second feedback loop, represented by a gain factor and a time delay $K_j e^{-\tau_j s}$, operating in the frequency region of interest. Therefore, the closed-loop neuromuscular system has third-order dynamics plus a zero due to the spindle pole in the feedback loop. Data obtained by McRuer *et al.* (1974) indicate that the muscle/manipulator dynamics for rudder pedals and hand manipulators are similar in form and numerically, despite the difference in limb size and function. Van Paasen *et al.* (2004) suggested an extension to the model in which the manipulator and the human arm are not unique blocks anymore, but their interaction is considered. Such kind of analysis is useful in application scenarios in which the human subject is operating while subjected to accelerations (i.e., in a moving vehicle) or is using active manipulators in which an active servo element is used to provide feedback from the controlled system or from other sources in the environment.

Research efforts have also been directed toward analyzing the relationship of human performance in manipulative tasks and muscle fatigue dynamics. An example of such muscle fatigue and recovery models is proposed by Fayazi *et al.* (2013), which links the maximum voluntary contraction (MVC) to the output isometric force in a cycling application. Liang *et al.* (2009) proposed a model in which the muscle capacity after a certain number of contractions is evaluated and put in relation to the external load force. A further extension of this analysis also considered the relationship of MCV with brain effort, distinguishing between fatigued and non-fatigued motor units (Liu *et al.*, 2002). Other research activities that rely on different types of modeling techniques were used, such as by Uppal and Vaz (2022) and Mishra and Vaz (2017), where a bond graph mathematical model was used to describe biomechanical characteristics of upper limb tendons during grasping. Vaz *et al.* (2015) used bond graphs to describe the extensor mechanism of a finger, being represented as deformable strings and assumed to pass through hooks fixed at predetermined points on rigid phalanges.

However, all the descriptive models of neuromuscular dynamics in the literature operate in high frequencies. This consideration leads to the fact that when neuromuscular dynamics are considered as an element of a more general control model, such as the ones that will be analyzed in the following section, often only their low-frequency effects are taken into account. Such effects can be simplified as a delay element.

2.3 Sensory Dynamics

Although all sensory organs are well known individually, as well as their dynamic behavior, their joint role with the CNS in perception has been of interest for further investigation quite recently (Lone and Cooke, 2014). One example that motivates recent interest in sensory dynamics modeling is the operator's disorientation. Although it is commonly taken for granted that reality can be accurately perceived, situations in which a human is subject to continuous rotations may lead to spatial disorientation. Spatial disorientation occurs when the human operator fails to perceive his position, motion, or attitude correctly. Research works in this field date back to the later decades of the eighteenth century, initiated by Ernst Mach and his colleagues with a study on vestibular and acoustic perception. However, true progress was achieved only after almost a century, in the 1990s, when mathematical modeling of spatial disorientation was proposed (Tsang and Vidulich, 2002).

2.3.1 *Visual system*

The human visual system is our sensory system's primary information source. It was vastly studied as a mathematical model with reference to computer vision techniques or the development of simulators for vehicles or aircraft. The importance of this sensory modality in the last-mentioned application scenario is confirmed by the fact that such a simulator very often relies on a fixed-base structure, thus not stimulating the vestibular system. Human vision can operate mainly in two modalities: ambient or focal mode.

The ambient mode mainly intervenes with humans' spatial orientation capabilities and relies on several inputs from the central and peripheral vision systems, such as motion, perspective, texture, and brightness gradients. The most interesting characteristic of this visual mode is its capability to subconsciously process if there is any disturbance over the aforementioned input signals and provide a stable perception covering a large spatial range. This is done by sending low-frequency, robust signals to the CNS, which uses this information to determine spatial orientation. Conversely, other sensory systems provide high-frequency transient signals to help stabilize the perceived surrounding environment immediately after motion. The ambient mode is very useful for perceiving distance and the angle between the operator's plane and the ground (i.e., slant). Lone and Cooke well described the possible sources contributing to the spatial disorientation (SD) of a pilot guiding a vehicle (Lone and Cooke, 2014). Visual perception of both slant and splay angles, being respectively the relative orientation in the vertical and horizontal plane, can lead to misjudgments about humans' estimation of the controlled element's actual position.

The focal mode is linked to object identification and relies mainly on binocular signals coming from the central visual field. It provides very detailed information at high spatial frequencies and is usually represented in conscious states (Previc and Ercoline, 2004). Figure 2.3 represents the visual perception model that Hess proposed by Curry *et al.* (1976), providing a simple way to model visual observation. Saturation limits can be set by considering two times the value of the variance of the random number generator input. Consequently, the variance determines visual signal quality and is related to the relationship between usable cue environment (UCE) and visual cue rating (VCR) (Schmidt and Bacon, 1983). Its value can be selected between the following ranges:

$$0 < \delta_{\text{VIS}}^2 < 0.1 \quad \text{if} \quad \text{UCE} = 1, \tag{2.4}$$

$$0.1 < \delta_{\text{VIS}}^2 < 0.2 \quad \text{if} \quad \text{UCE} = 2, \tag{2.5}$$

$$0.2 < \delta_{\text{VIS}}^2 < 0.3 \quad \text{if} \quad \text{UCE} = 3. \tag{2.6}$$

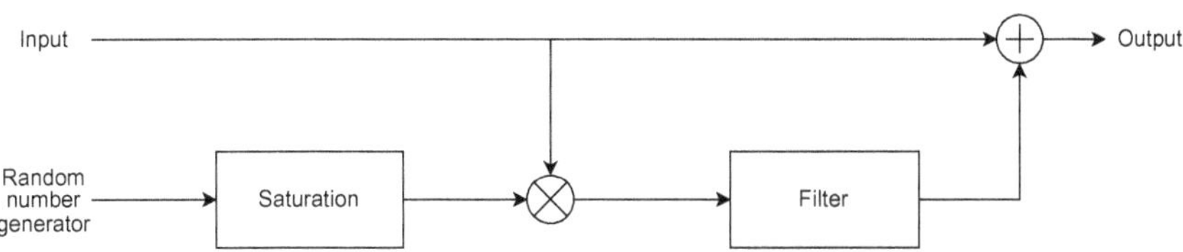

Fig. 2.3. Visual cue perception model proposed by Curry (1976).

This parameter has been extended to task-dependent variance related to vision with multiple axes:

$$\delta^2_{\text{task}} = \begin{cases} 0.01n & \text{if } n > 1, \\ 0 & \text{if } n = 0, \end{cases} \tag{2.7}$$

where n is the number of controlled axes. The two terms can be incorporated into the following factor:

$$f = 1 + 10(\delta^2_{\text{VIS}} + \delta^2_{\text{task}}). \tag{2.8}$$

Along with the global view of vision modalities, human vision was also studied in relation to object tracking, particularly in computer vision and image processing fields. Nguyen *et al.* (2020) recently developed a tracking model that utilizes the spatial-temporal context information to increase tracking accuracy level. Further improvements in visual tracking research were the widespread adoption of discriminative learning methods (Henriques *et al.*, 2015). These types of classifiers are tasked with distinguishing between the target and surrounding environment, often used in order to ensure target tracking in the presence of occlusions (Chen *et al.*, 2020). The latest trend in the field is multiple object tracking (Luo *et al.*, 2021); the challenge, in this case, lies in locating multiple objects, maintaining their identities, and yielding their individual trajectories given an input video (in the case of computer vision applications). Object to label in this case can be pedestrians (Kumar and Ghosh, 2022) or vehicles (Roy *et al.*, 2022) in the road safety management field.

2.3.2 *Vestibular system*

The vestibular system is responsible for human equilibrium, postural control, and the proprioceptive sense of body motion. Anatomically, it is housed in the inner ear and can be divided into semicircular and otolith canals. The otoliths perceive a sense of tilt and force, while semicircular canals help provide the sense of angular acceleration. Accurate analysis and estimation of their dynamic response have been crucial for human perception modeling when interacting with any mobile-controlled machine.

Angular motion, characterized by low amplitude, is limited by inherent thresholds, which are a function of the stimulus magnitude and its duration. Mulder's law describes angular accelerations with a duration inferior to 10 s as the product of angular acceleration, and its duration is approximately equal to 2.5 deg/s. This means that a weaker acceleration requires more time to be perceived from vestibular canals. In the aerospace domain, experimental studies on human sensory thresholds for angular velocities and accelerations characterized by prolonged duration have been done (Heerspink *et al.*, 2005). It is suggested that such thresholds can vary depending on the nature of the controlled elements. For example, flight can be slightly higher with respect to a car due to more stress and, consequently, the pilot's attention level and allocation. The pilot's experience and training contrast this effect; in this case, the human has an accurate internal model of the machine's dynamics, allowing him to have a certain degree of knowledge in advance and lower the threshold. In summary, the workload, stress level, and training level strongly impact the human sensing abilities of a rotational motion. Being very difficult parameters to quantify, an accurate model of threshold dynamic variation is very hard to obtain. The main role of otoliths relies on the sense of linear accelerations and vertical motions, with threshold levels of 0.1 g and 2 deg. Otolith canals cannot differentiate between acceleration caused by gravity and other linear accelerations. The sensed motion should always be considered as an apparent vertical motion since there is no difference in the way humans can perceive tilt and linear accelerations.

Further attempts to model the vestibular system led to the development of Hosman's descriptive model, whose main purpose was to integrate visual and vestibular dynamics, which will be better discussed in Section 2.5.

2.3.3 *Proprioceptive systems*

Among the first senses to develop in a human being are, without doubt, tactile and proprioceptive senses since they are mandatory for determining the gravity vector and consequently developing the necessary anti-gravity group of muscles, which allow us to walk.

Proprioception, which is also called kinesthesis, refers to such a sensory modality which uses muscle spindles to determine the position of the body and limbs of the subject, as well as their movements and the joint torques required to start a motion or maintain a steady position against resistive loads. In human–machine interaction, the role of tactile and proprioceptive systems is linked mostly to the force and pressure feedback that the operator has due to the physical contact with an aircraft inceptor, a vehicle steering wheel, or a robot's end-effector.

Pressure receptors are located within the skin all over the human body and are of primary interest in the development of modern haptic feedback devices, which are able for example to provide information about a surface texture belonging to an unknown external environment in teleoperation frameworks.

The modeling of this system is difficult because of the huge number of physical stimuli which trigger its response. Classical factors that trigger an output of the proprioceptive system directed at the CNS are, for instance, relative linear and angular velocities and muscle tension and its orientation with respect to the gravity vector. Usually, all these inputs are sensed and elaborated simultaneously. Since, as is evident, some of these factors also stimulate other sensory systems, the CNS combines in the case of conflict the multiple received sensory information to develop its own proprioceptive sense. For this reason, proprioception cannot be seen as a unique sensorial system like visual and vestibular ones but should be considered more like a sense developed as a combination of different sources of information. Hess provided a transfer-function representation of proprioceptive dynamics in his "structural model," which will be discussed in detail in Section 2.5.

2.3.4 *Inter-sensory models*

A relatively recent field of study is the human cognition associated with multi-sensory stimuli. The first steps in this context were provided, once again, by neuroscientists such as Halligan (1999) and Lotto (2011), even if the considered interaction between senses was limited and focused on forms of synesthesia and goes toward a brain function associated with a high level of complexity.

Multi-sensorial perception is mostly modeled as a simple linear summation of inputs, or as a weighted sum, with an almost arbitrary selection of weights. The most complex developed model is Hosman's descriptive model, which has a nonlinear combination of visual and vestibular stimuli.

Telban and Cardullo proposed a model able to capture the perception of rotational motion, parametrized in order to match latencies experimentally observed by Berthoz *et al.* (1975) and Fernandez and Goldberg (1971). This model is able to analyze inputs coming from peripheral and central visual fields, as well as vestibular inputs. The rotational perception model, which is represented in Figure 2.4, provides the computation of the perceived angular velocity, given the actual inputs coming from the two considered sensory systems, where the semicircular canals represent the vestibular one. A similar model is the translational

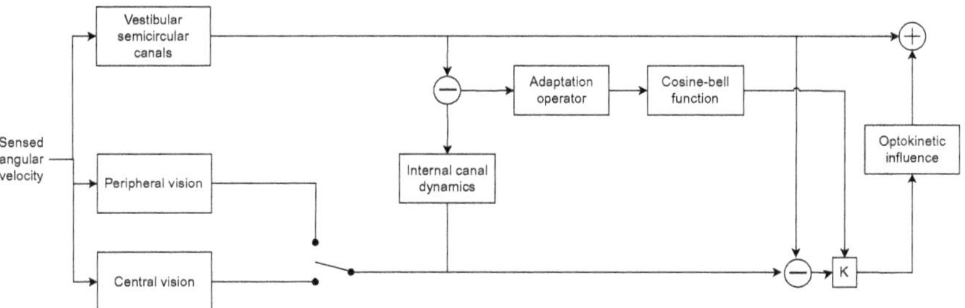

Fig. 2.4. Telban and Cardullo's rotational perception model (Telban *et al.*, 2000).

perception model, in which the perceived velocity and acceleration are obtained given the actual specific force. Vestibular dynamics in the latter case are represented by the otolith canals, which respond to specific force stimuli, while the visual system processes the velocity information, which is mathematically represented as an integrated acceleration. Back to the rotational perception model, peripheral and central vision are considered time delays, respectively set to 90 ms and 150 ms (Hosman and Stassen, 1999). Further psycho-physical experiments have provided evidence that visual perception of self-movements can induce an artificial vestibular response (Lone and Cooke, 2014). The opposite process can also happen, even if to a limited degree. This model's main feature is the capability to represent such influenced the estimates of self-movements. Optokinetic influence components provide both a nonlinear gain element and a first-order low-pass filter. The gain element is able to represent the weight given to vestibular and visual perceptions and is calculated from a cosine-bell function, which links it to the difference between them. The low-pass filter models the semicircular and otolith canals, implicitly assuming that the CNS compares the visual stimulus with its estimation of vestibular response. For what concerns vestibular models, a certain degree of correspondence can be noticed between the model proposed by Fernandez *et al.* (1971), Telban *et al.* (1971), and Hosman (1999). In all of these models, the otolith organs respond to a specific force, defined as

$$f = \hat{g} - a_h. \tag{2.9}$$

Here, $\hat{g}$ represents the local gravitational force vector, while a_h is the acceleration of the head of the human operator with respect to a fixed reference frame. Assuming, for the sake of simplicity, that the operator's head is aligned with the fixed frame axes, it is possible to obtain the transfer function between sensed and actual force:

$$\frac{\hat{f}(s)}{f(s)} = \frac{0.4(13.2s + 1)}{(5.33s + 1)(0.66s + 1)}. \tag{2.10}$$

While for what concerns the perceived and actual angular rotations, the transfer function can be expressed as

$$\frac{\hat{\omega}(s)}{\omega(s)} = \frac{456s^2}{(5.7s + 1)(80s + 1)}, \tag{2.11}$$

providing a reliable representation of vestibular canal dynamics. The adaptation operator element indicates the maximum time for which it is possible to have a conflict between vestibular and visual inputs. This is done by relating the inter-cue error to the washed-out error:

$$\frac{e_w(s)}{|e(s)|} = \frac{\tau_w s}{\tau_w s + 1}, \tag{2.12}$$

where τ_w represents a time constant. Simulation results of the model (Lone and Cooke, 2014) showed that the model is capable of representing the difference between the transient nature of the vestibular response and the constant presence of visual stimuli in human motion perception.

2.4 Human–Machine Control Models

In this section, human models will be discussed from a global point of view, with a control-theory approach. In the presented models, the human–machine system is described with a task-dependent approach, typical of control science. Here, significant variations can be observed regarding the contribution of physiological structures described in the previous sections as subsystems in the overall model, as well as the abstraction level of their mathematical representation.

2.4.1 *McRuer's crossover model*

From the very beginning of the studies in this field, McRuer *et al.* (1965) analyzed human control action in compensatory tasks by randomly changing the target reference trajectory which the human had to follow. The result was one of the most common and simple examples of a human control model, McRuer's crossover (CO) model, also known as the quasi-linear model. The quasi-linear model hints at how humans adapt to different plants to elicit stable and effective control responses.

It can be convinced that such a model exhibits the behavioral invariance of the human in its adaptation to the controlled machine, offering a consistent human–machine behavior where the functional block diagram can be described as a simple compensatory manual control system. Due to its simplicity, the model's proposed aim is to avoid common problems related to higher complexity systems.

It was observed that when an external disturbance is introduced in the system, measured human operator responses were different for different transfer functions of the controlled plant, but the combined human–machine behavior is approximately the same for all the experiments. The following equation can describe the transfer function of the combined human–machine system:

$$Y_p(j\omega)Y_c(j\omega) = \frac{\omega_c}{j\omega}e^{-i\omega\tau}, \tag{2.13}$$

where ω_c is the CO frequency of the system, and τ is the overall delay in human response. Such an equation indicates that the behavior of the human–machine complex can be described as a simple integrator and a delay in the CO region. If we isolate the human controller from the controlled element, the whole system relating to a voluntary motion can be simplified into three components: a linear controller inside the brain, a neuromuscular dynamic, and a reaction time delay.

After the learning phase of the machine dynamics is sufficiently finished, the human can be considered as a simple feedback controller which moves the controlled element to the target position, in case of a point-to-point (PTP) task, by watching the reference target point.

The neuromuscular dynamics, as said, can often be approximated by a first-order lag, as demonstrated also by Phatak *et al.* (1976), and the simplest human controller was modeled as a PD controller by Ragazzini (1948); therefore, the human transfer function $Y_p(s)$ can be described as follows:

$$Y_p(s) = K_p\frac{\tau_L s + 1}{\tau_I s + 1}e^{-\tau_e s}, \tag{2.14}$$

where the parameter K_p is the pilot's gain, τ_L is the lead time constant, τ_I is the lag time constant, and τ_e is the pilot's reaction time delay. The parameter selection is carried out by using the adjustment rules. According to the model, the reaction time delay should be constant for each human subject (Suzuki and Furuta, 2012), with small variabilities due to task and environmental variables.

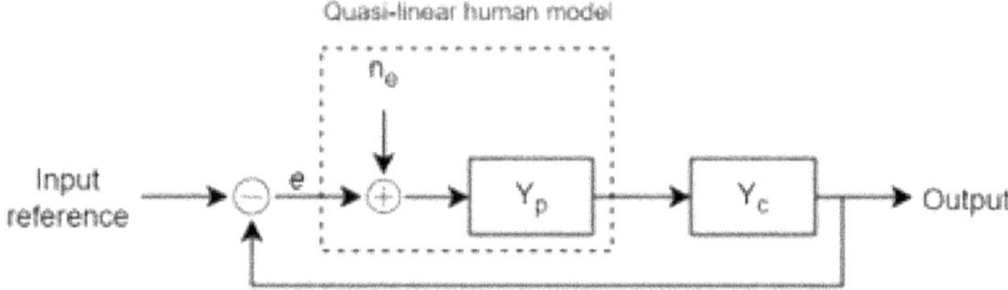

Fig. 2.5. Simple feedback structure for a human–machine complex in manipulative compensatory tasks according to crossover model.

In the quasi-linear model represented in Figure 2.5, McRuer introduced the remnant noise term n_e to account for the non-stationary effects of the pilot behavior. The remnant was described as a random process, linearly uncorrelated with the control input. Normally, the remnant is related to the error signal $e(t)$. Therefore, it is here considered as the observation noise.

Although originally born to describe pilot dynamics, the CO model has become a benchmark in human control models for a variety of applications and controlled elements. For instance, a generalization of the CO model is proposed by Martínez-García *et al.* (2017), characterizing the human control of systems with both integer and fractional-order plant dynamics. Or, in teleoperated surgical robotic systems (Takacs *et al.*, 2005), it can be used for a detailed characterization of operational delay in order to improve control precision.

2.4.2 *Optimal control model*

The optimal control model has been developed by Baron *et al.* (1970), Kleinman *et al.* (1970), and Wierenga (1969) in the first place, in consequence of the advances in optimal control theory, which can be observed between the 1970s and the 1980s. The main concept behind it is that after a certain level of training and motivation, a human operator can control a machine in an optimal manner, even if it remains subject to physical and psychological limitations. The first observable difference from the CO model is that the optimal control model was expressed using state space variables, which makes it easier to extend human–machine analysis to multi-loop control tasks.

Figure 2.6(a) shows the first simple version of the optimal control model (OCM). If we consider a visual input reference y, the first process to consider is the pilot's reactive time delay, while the signal y_p is the perceived input

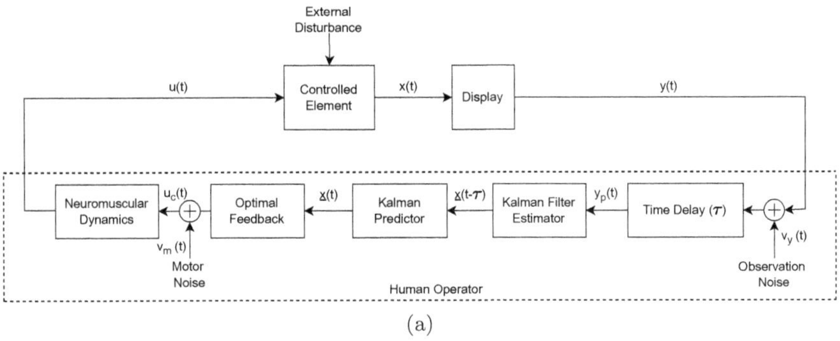

(a)

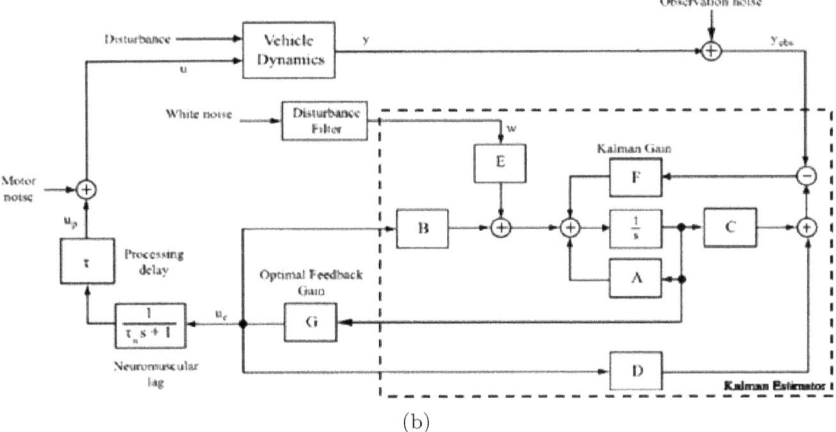

(b)

Fig. 2.6. (a) Optimal control model of the human operator, as defined by Kleinman *et al.* (1970). (b) Modified version of the optimal control model for pilot–vehicle dynamics (Schmidt, 1985).

signal, namely the internal image of the actual input y in the CNS of the human pilot.

Neuromuscular dynamics make the pilot execute the optimal control and can be expressed by a first-order lag $\frac{1}{\tau_N s+1}$. Moreover, u is the output human's control action, x is the internal state vector of the controlled element, w is an external disturbance, and y is the vector containing external sensed measurements. The elements of estimation and decision consist of the following:

- Kalman filter, which is used to model a human's ability to deduce a system state from perceived information;

- Kalman predictor, which represents the compensation for inherent time delay;
- optimal feedback, which builds optimal control u_c based on y_p input.

All of these elements require a model of the controlled machine and, therefore, can be considered as the human's internal representation of the machine's dynamics, with all the deriving linearization processes and other psychological limitations (Innocenti, 1988).

The order of the model is dependent on the human training level and expertise and can include limited models of actuation systems. The validation of OCM can only be performed with a black-box approach, comparing its output to real human control outputs. In systems in which a single input is considered, the performance index reflecting human control strategy can be expressed as the following quadratic cost function:

$$J = \lim_{T \to \infty} \frac{1}{T} E \left\{ \frac{1}{2} \int_0^T (y^T Q y + g\dot{u}^2)dt \right\}, \qquad (2.15)$$

where E is the expected value, Q is the weight coefficients matrix, while g is a real weight coefficient chosen so that $g > 0$. In manual control compensatory experiments, the element $y^T Q y$ is usually set in order to minimize mean squared error; here, the determination of Q was performed only by using empirical methods. Moreover, the factor $g\dot{u}^2$ sets a superior threshold on the total energy which can be used in a control task. The inclusion of such a term into the cost functional results in a first-order lag, which is often associated with neuromuscular dynamics (Baron *et al.*, 1970); in fact, given the other parameters of the model, there is a direct proportionality between g and τ_N. Moreover, along with empirical methods, the values of model parameters were also numerically computed by identifying the OCM model in the works of Phatak *et al.* (1976) and Doman (1999).

However, the model accuracy when matching real data has not been significantly improved with respect to traditional control models, which indicates a certain over-parameterization. For this reason, OCM has been improved to the form of a modified optimal control model (MOCM) developed by Schmidt (1985) and represented in Figure 2.6. Another modification developed in parallel led to the fixed-order OCM (Doman and Anderson, 2000). Both of them offer transfer function representations with frequency domain analysis. This work can be considered as the transition phase between the classical frequency-domain and more recent time-domain approaches. However, the simplification process which motivated their development was contrasted by their complexity. Evidence of this concept is that Schmidt himself chose to use the full parameter model for capturing the effect of aircraft elasticity in his human-in-the-loop simulation and analysis (Baron and Kleinman, 1969).

A revised optimal control model (ROCM) of a pilot, which is based on the aforementioned modified version, was also presented by Wang *et al.* (2008). This model was later extended for the analysis of different aspects in further research works: human decision-making (Gai and Curry, 1976), its monitoring behavior (Kleinman and Killingsworth, 1974), the execution of multiloop tasks (Hess, 2008), and multimodality (Hess, 2010), where models of semicircular and otolith canals of the vestibular apparatus were provided. Overall, the optimal control model was used to solve a number of applicative control issues, mainly in pilot–aircraft interaction tasks, such as the prediction of flying qualities (Schmidt and Bacon, 1983) and the use of such predictive capability in refueling tasks (Efremov *et al.*, 1998). Other research activities were dedicated to the definition of a relationship between Cooper–Harper ratings and the cost function expressed by equation (2.15) in both single- and multi-loop tasks (Edkins, 1994; Hess, 1977). Moreover, the optimal control model was used for many other applied research activities, such as the simulation of pilot control strategy when encountering wake vortex (Schönfeld, 2010), the assessment of loads in airframe flights (Lone, 2013; Lone and Cooke, 2012), or the investigation of display dynamics on the control loop in order to obtain relationships between display types (Hess, 1972; Levison, 1989) and ratings of the human operator (Doman, 1999), as well as many other research works (Gawthrop *et al.*, 2011; Jin and Orosz, 2018; Johannsen and Rouse, 1979; Mainprice *et al.*, 2015; Menner *et al.*, 2019; Pattipati *et al.*, 1983).

2.4.3 *Structural model*

In spite of the successes of linear models described in the previous sections in investigating the relationship between human control dynamics and handling qualities, as well as their application to analysis/design problems, they both lack an accurate description of the underlying physiological control structure contributing to human pilot dynamics (Hess, 1979). Moreover, further research done in the same period showed that when the difficulty of a control task increases, the human control behavior becomes highly nonlinear.

Hess's studies were motivated by two main observations: (i) human operator control strategies often seemed to result in discrete or impulsive motions; and (ii) such experimental evidence was not linked to any feature of the classical linear control models. The main assumption of such investigations was that the operator, when associated contemporaneously with a high-order-dynamics vehicle and a difficult task, tends to reduce the overall complexity load associated with the time integration of multiple sensory inputs (Hess, 1979). This simplification is done by simply adopting a nonlinear strategy relying on a limited number of parameters (rather than a linear strategy associated with a high complexity level and the number of parameters).

The first development of these assumptions can be found in the isomorphic model (Smith, 1975), which can be considered the father of the successive structural model described by Hess. The main idea relying on them is to better describe human signal processing by determining feedback paths from the sensory modalities involved in perception and motor control. The human equalization process, namely the human's "proprioceptive" feedback, occurs through this simulated feedback path, whose parameters were tuned to match the performance of the quasi-linear model near the CO region (Hess, 1980a, 1990a).

Figure 2.7 shows a block scheme representation of the structural model. Here, the element Y_{d_e} is the transfer function of visual dynamics when perceiving its input signal from a display. Moreover, n_u represents the remnant noise and, as in the quasi-linear model, is considered as an observation noise (and, therefore, put in the human's output), while d is an external disturbance that acts on the controlled element. The parameters K_e and $K_{\dot{e}}$ are the gains of the central processing stage, while τ_0 and τ_1 represent the correspondent time delays. In this model, the pulsing logic

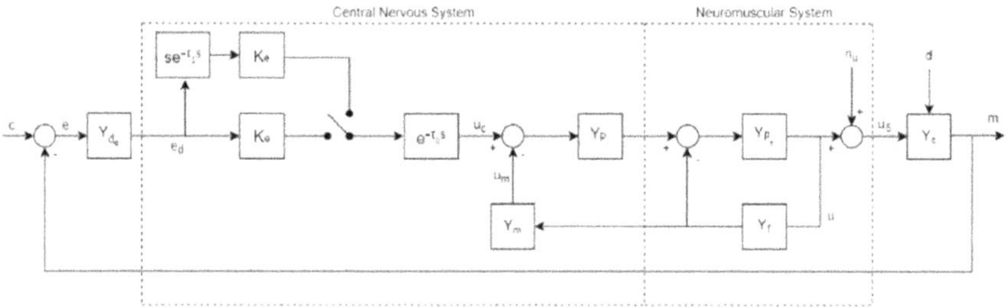

Fig. 2.7.   Hess's structural model of an adaptive human pilot.

$Y_{p_1} = 1$ (Magdaleno and Mc Ruer, 1971) and element

$$Y_m = \frac{K_m}{(s + 1/T_m)(k - 1)}$$

describes the aforementioned pilot inner-loop feedback. The key aspect is the selection of parameter k, which can be interpreted as the pilot's internal model of the controlled element dynamics and reflects the adaptive characteristics of the human pilot. It will mainly depend upon its transfer function around the CO frequency.

In the CO region, it becomes $Y_m \propto sY_c$. Therefore, the following general considerations can be made:

- $k = 0$: the controlled element is a constant;
- $k = 1$: the controlled element is an integrator;
- $k = 2$: the controlled element is a square integrator.

The representation of the pilot neuromuscular system includes both front and feedback channels. The describing function

$$Y_{p_n} = \frac{\omega_n^2}{s^2 + 2\xi_n\omega_n + \omega_n^2}$$

represents the open-loop dynamics of the limb which is driving the manipulator, while $Y_f = \frac{K_f}{s+1/T_f}$ represents the muscle spindles. After its early definition, the structural model was modified, extended, and applied to different scenarios. For instance, the simple application of a structural model in a tracking and regulation task resulted in a motion cue model (Hess, 1990b). Further experimental activities were directed toward the determination of time delay effects in manual control systems dynamics. In the subsequent analytical work, changes in human equalization performance were observed because of such time delays (Hess, 1984).

Following several modifications done to the original version of the structural model, Hess developed his revised model in 1997, willing to include the effects of the pilot's neuromuscular system characteristics in the aircraft control process, along with its ability to perceive forces. Further extensions of the structural models are intended to better specify motion and force feedback (George, 2008). Still, in the aeronautic domain, Heffley (2010) developed the task–pilot–vehicle (TPV) model, which is a simple extension of the structural model in tracking and regulating tasks applied to fast system design. These modifications proved to have a good match with the real data of human describing function (Efremov and Tjaglik, 2011).

Such kinds of modified versions of the model were used in many practical control problems, such as the design of a predictive display that also considers motion cues (Efremov and Tjaglik, 2011) or the development of an analytical method to assess a pilot's fidelity in flight simulators (Hess and Malsbury, 1991), including multi-loop tasks (Hess *et al.*, 1993; Hess and Siwakosit, 2001). The procedure which led to the pilot's assessment was performed in Zeyada and Hess (2003), where a flight was simulated using a six-degree-of-freedom controller of a rotary wing aircraft, which was executing a vertical maneuver. A structural model of the human was also used to explore the closed-loop nature of a pilot's control behavior when determining a target direction and the analysis of the characteristics of pedal feedback to the pilot (Hess, 2005). Finally, evaluations of the qualities of aircraft handling were proposed by Hess (1989) and Hess and Sunyoto (1985). Their prediction, along with the estimation of pilot-induced oscillations, was applied to different controlled elements in the following years (Afloare and Ionita, 2014; Grant *et al.*, 2015; Hess and Joyce, 2013).

2.4.4 *Descriptive model*

In Europe during the 1970s, technological development such as the fly-by-wire (FBW) aircraft again led to research interest in man–machine interaction for aeronautic scientists. Researchers started to consider their human modeling effort as a means to better understand flying handling qualities and performance (Gibson, 1997; Padfield, 1998) and to improve the accuracy of flight simulation. For this purpose, better integration of visual and vestibular systems into the control model seemed necessary. Later, in the 1990s, Hosman led extensive research efforts with the aim of understanding the influence of the visual and vestibular systems on human perception and, consequently, on its control behavior (Hosman, 1996; Hosman and Stassen, 1999). This investigation resulted in the definition of the descriptive model, for which Hosman presented different experimental works in which the case study used was a pilot's landing maneuver with an aircraft using a moving-base flight simulator (Hosman *et al.*, 2005). There, systematic variations of sensory inputs were the base that led to the definition of the descriptive model (Hosman and Van der Vaart, 1981). Results were applied in closed-loop control tasks where the human was considered as a single-channel information processor with multiple inputs from the sensory systems.

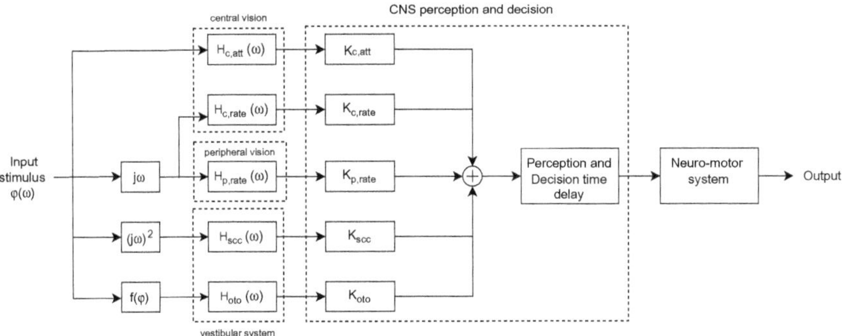

Fig. 2.8. Hosman's descriptive model of human control behavior (1996).

The descriptive model, represented in Figure 2.8, has a multimodality structure, reflecting physiological subsystems that link the states of the controlled element to his perceived states. Here, the element of visual perception of displacement is expressed in the time delay $H_{\mathrm{aat}}(s) = e^{-\tau_{\mathrm{att}} s}$, where the attitude perception delay τ_{att} resulted in being around 50 ms (Heerspink *et al.*, 2005). The visual perception of a velocity is $H_{\mathrm{rate}}(s) = e^{-\tau_{\mathrm{rate}} s}$, where the delay parameter varies in the case of central or peripheral vision. In particular, the peripheral system is able to sense only rates; therefore, its dynamic is described by the second equation, but with a shorter time delay (60 ms) with respect to the one measured in the central vision system (110 ms) (Hosman and Stassen, 1999). For example, the delay referred to by McRuer in his CO model is the sum of the delay associated with the detection of the stimulus in the eye and the one associated with the information processing.

For what concerns the vestibular system, both the semi-circular and otholit canals are modeled, respectively, as an over-damped torsion pendulum and an accelerometer with over-damped mass–spring–damper characteristics. They both can be represented by second-order differential equations, having the following transfer functions:

$$H_{\mathrm{scc}}(s) = \frac{(1 + \tau_L s)}{(1 + \tau_1 s)(1 + \tau_2 s)},$$

$$H_{\mathrm{oto}}(s) = \frac{(1 + \tau_n s)}{(1 + \tau_a s)(1 + \tau_b s)}. \tag{2.16}$$

In his model, Hosman assumed that tactile and proprioceptive senses were implicitly considered within the vestibular dynamics model. Moreover, the descriptive model assumes that processing in the perception and decision stages by the CNS and neuromuscular dynamics can be combined and represented by a unique "information processing" transfer function:

$$H_{\mathrm{ip}}(s) = K_{\mathrm{ip}}e^{-\tau_{\mathrm{ip}}s} . \qquad (2.17)$$

Here, the CNS contributes both to the gain and delay elements, while the neuromuscular dynamics low-frequency effects are approximated by only a delay factor; thus, τ_{ip} is the sum of delay contributions of both factors. The function f converts the displacement caused by the input stimulus to a specific force output. The descriptive model has been applied in numerous studies in the transport engineering field, focused on the identification of human pilot's dynamics, such as the implementation of optimal forcing functions for identifying human model parameters (Zaal *et al.*, 2008), the investigation of the use of visual information by the operator while controlling an aircraft (Mulder *et al.*, 2005), or the study of the influence of translational movements on the pilot's performance and perception (Grant *et al.*, 2005; Groen *et al.*, 2007).

2.4.5 *Biodynamic models*

The biodynamic models were developed to represent the effects of the body dynamics on the human's desired control input in a situation in which it is subject to an accelerating environment. This way, the effects of this kind of motion on human health, comfort, or performance can be predicted (Griffin, 2001). Human control actions can be divided into two general categories: voluntary and involuntary (Masarati *et al.*, 2015). The models studied in the previous sections were all part of the first category. All of them focused on translating into mathematical expressions the process which converts an idealized voluntary human action into an actual control action. This process takes into account aspects such as bandwidth limitations and the delays in human dynamics, considered in relation to rigid-body movements and human-induced oscillations.

An alternative kind of modeling technique aims to describe the involuntary human actions, being the direct consequences of the vibrations coming from the environment, which are filtered by the human's body and become an involuntary input into its control system (Dieterich *et al.*, 2008; Pavel *et al.*, 2013; Venrooij *et al.*, 2014a, 2014b).

The modeling of biomechanics can be categorized into three types: continuum, discrete, and lumped parameter models. The main difference between them is the way in which the spine is modeled.

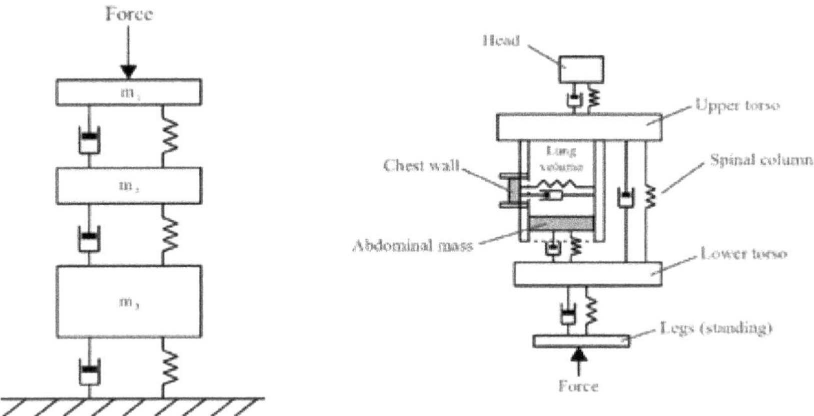

Fig. 2.9. Examples of discrete (left) and lumped parameter (right) model structures (Gribble and Ostry, 1999).

In continuum models, the spine is considered a flexible beam; its responses to vertical accelerations were studied by Griffin *et al.* (2001). In discrete models, shown on the left side of Figure 2.9, the spine is described as a series of interconnected mass–spring–damper elements. The dynamic response can be studied by determining its equation of motion. With respect to the continuum modeling approach, discrete models succeed in representing the human body as a composite of elements (i.e., organs), each one with its own resonance frequency.

In lumped parameter models, depicted on the right side of Figure 2.9, the body itself is modeled as an equivalent mass–spring–damper system. Of course, this approach causes models of this type to have only one or two degrees of freedom. Moreover, the efficacy of its analysis is limited to the response to vertical stimuli (Kitazaki and Griffin, 1997a) and is not able to capture the complex dynamics of the human body, as evidenced by Sirouspour *et al.* (2003) when attempting to model lateral dynamics of a seated subject. This kind of approach was able to avoid instabilities and cancel dynamic feedthrough (Sövényi and Gillespie, 2007). Further efforts in biodynamic modeling are the whole-body transmissibility (Kitazaki and Griffin, 1997b; Matsumoto and Griffin, 1998), which is the ratio between the vibration measured in a certain point of interest and the base vibration (both are functions of frequency).

The area which produced the most research activities in this field was, once again, the aerospace domain. Levison and Harrah (1977) implemented a simulated biodynamic model able to predict both the human dynamic response and tracking performance in vibrating environments, allowing gathering data for whole-body vibration. Based on this model, further experiments were conducted to simulate the transmissibility of vertical base vibrations to lateral and roll accelerations (Hoehne, 1999; Kitazaki and Griffin, 1997a; Matsumoto and Griffin, 2001). One of the first important studies on biomechanical effects associated with human–manipulator interaction in high frequencies was carried out by Johnston *et al.* (1988).

Modern technological developments of this century, characterized by higher speed of transport systems, motivated further research investigation on the effect of structural vibrations of civil transports (Raney *et al.*, 2002) and supersonic aircraft (Raney *et al.*, 2001). Later, muscular damping and stiffness parameters were studied in relation to the urgency of the performed task (Venrooij *et al.*, 2009), suggesting that during tasks that are perceived with a higher urgency level, the body stiffness increases, such as the pilot's grip force. Moreover, Coe *et al.* (2009) defined a three-dimensional human body model which relied on collected experimental data, while Masarati *et al.* (2013) investigated the error between the intended and actual control action of the human and derived it in relation to biomechanical models of the limbs.

2.5　Final Considerations

The study of models representing human control behavior when interacting with a controlled machine has played a crucial role in many engineering fields. Emerging technical challenges, such as the development of new robots, aircraft, and vehicles, or the spread of advanced simulation frameworks, have motivated through the years the rise of more and more complex mathematical representations. Along with the rise of the application's complexity, the abstraction level of human behavior skill-based description has also increased, making it more difficult for the objective to include accurate functional modeling of the involved physiological structures. For this reason, the parallel research efforts which were performed in motor control theory and sensory feedback description were reviewed in this work. The open discussions in cognitive-science-related research activities were also detailed, concerning the presence of internal models of external dynamics with whom a human is interacting in their CNS. Correspondent differences in the functional description of the motor apparatus were also identified.

For what concerns the review of human sensory modeling, there were included subjects such as visual perception errors and their modeling, as well as vestibular and proprioceptive sensory dynamics.

Neuromuscular dynamics representations completed the discussion on the control theory representation of involved physiological districts, considering modeling efforts of internal feedback loops provided by tendons and spindles, forward muscular activation dynamics, and their approximation in the frequency range of a manipulative control task.

Passing to a task-based description of the human–machine complex, the aforementioned physiological dynamics were included in more general control structures in which different higher-level human features were also represented. Thus, the human capability to behave in an optimal manner after a certain level of training was represented in the optimal control model, while its adaptability to the machine dynamics is well captured by the classical CO model or by Hess's structural model.

The degree of integration of underlying physiological processes and mission-based strategies within these models is indeed really variable, being the advantage of approaches of the structural model and Hosman's descriptive model with respect to previous classical quasi-linear approaches. The ongoing development of the higher-level representation of human control strategies and decision-making, which includes techniques taken from robust control theory, uncertainty propagation, and probabilistic methods, is motivated by the increasing complexity of the application tasks. Therefore, maintaining a bond within the developed models from both a physiological and a control-theory point of view will be increasingly challenging.

Having focused on the linear models, in accordance with the relevant research efforts of the last few years directed at the description of nonlinear dynamics in man–machine interaction, the reader could be prompted to face this topic starting from the simple models overviewed, which have the advantage that they could be handled by using the classical approach of linear system theory and automatic control methods.

Despite the limitations, these linear models showed good approximation capabilities and ease of use in practical applications; in the following chapter, we will present a novel linear modeling strategy applied to human–robot interaction.

Chapter 3

Precision Model for Human Reactive Delay Identification in Robotic Manipulation

Human–robot collaboration offers advantages in terms of flexibility in many industrial applications. New generations of intelligent collaborative robots are used without fences, usually in a shared space with human operators, which adds a mutual advantage to classical robotic cells. In complex human–robot collaboration scenarios characterized by continuous physical interaction, analyzing human behavior and control action is crucial to investigate as a base to build any predictive technique. The previous chapter showed how modeling the human control action when interacting with a controlled machine has become an independent research field involving multiple disciplines and approaches over the years. This chapter investigates and applies a linear precision model to human–robot interaction, focusing on identifying human reactive delay in a collaborative task.

3.1 Human-in-the-Loop Control

Several studies regarded human and robot performance evaluation in cooperative work. Fruggiero *et al.* (2020) evaluated the human factor in virtual scenarios using agent modeling with a discrete transition between blocks. Sadrfaridpour *et al.* (2016) built a trust model in which the performance of the robot and human operator in repetitive collaborative tasks is evaluated, considering the human muscular fatigue and recovery model.

Different modeling approaches have applications in human intention estimation and robot imitation algorithms. For example, Obo *et al.* (2015) proposed a human-like robot posture generation method based on a steady-state genetic algorithm (SSGA). Huber *et al.* (2008) tried to dynamically understand the joint action of groups of humans working together and transfer such behavioral patterns to human–robot interaction. Erlhagen *et al.* (2006) have developed high-level joint action strategies for human-human interaction. Such approaches either assume perfect rationality or smooth over human idiosyncrasies and noisy observations by providing general accounts and mathematical functions of human performance.

Recently, promising approaches to describe the information transfer relating external dynamics to the human controller have been proposed using convolutional neural networks (CNNs) or other machine learning techniques. Manjunatha *et al.* (2022) used a transfer learning approach based on a CNN with raw EMG input data to classify motor control difficulty in a human–robot collaborative scenario. Chen *et al.* (2022) used LSTM networks to learn the characteristics of strongly nonlinear external dynamics of Van der Pol and Lorenz systems. Mu *et al.* (2020) used a reinforcement learning algorithm coupled with two neural networks to implement an event-triggering dynamic strategy for partially unknown systems.

Concerning this kind of black-box approach, other model-based techniques achieved the same goal while developing a better description of the underlying physiological processes which determine the overall human control strategies. The underlying mathematical representation for human control action should be the same as the one used to describe a linear servomechanism (Levine, 2018). Specifically, a set of linear differential equations with constant coefficients, or, equivalently, a transfer function in the frequency domain. In 1965, McRuer *et al.* introduced the most famous example that follows such an approach. The model foresaw an integrator and a simple delay element to model the human operator adaptation during the interaction with the machine. The human action was studied with different controlled elements and represented a quasi-linear describing function. Kleinman (1969) developed the "optimal control model" (OCM). Its primary assumption is that a well-trained, well-motivated human behaves optimally while subject to psycho-physical limitations. Unlike the crossover model, the OCM is expressed in state-space variables, facilitating the extension of human–machine analysis to multi-loop control tasks. Here, a Kalman filter represented the human's ability to deduce a system state from perceived information. Concerning pilot aircraft control applications, Hess's structural model (1980b) proposed a detailed description of human perception processes and inner loop feedback. While Hosman's descriptive model (1999) studied the interaction between visual and vestibular inputs and their influence on the overall control strategy.

Research in this direction in the robotics domain often investigates the exploitation of collaborative robot applications using virtual environments. Cremer *et al.* (2019), for example, used a neuroadaptive controller in a collaborative application with a robotic arm to analyze the human control action. The target point was shown in a virtual environment and randomly changed for each trial. Whereas Suzuki and Furuta (2012) combined a similar virtual setup with a haptic device. The identified human control parameters were the input of a human intention estimator, able to adapt the impedance of the controlled device according to human actions.

However, these approaches do not investigate the effect of the learning process of external dynamics on human control delay. The classical quasi-linear model is the most suitable approach for this while focusing on the description of physiological background processes (Magdaleno and Mc Ruer, 1971; Scibilia *et al.*, 2022a; Suzuki and Furuta, 2012). The crossover model has often been used in literature for this purpose in the aeronautic domain, characterized by a high-frequency range that facilitates a frequency-domain analysis of the model.

The current work describes and applies such a modeling approach to different human–robot collaboration scenarios using a time-domain approach. The chapter is structured as follows: In the second section, the model theoretical background will be described in detail; in the third section, the proposed application scenario will be presented; in the fourth section, the obtained results will be analyzed and discussed; and in the final section, the system identification process implemented in simulation will be illustrated.

3.2   Precision Model of the Human–Robot Complex

Let us recall the McRuer's crossover (CO) model (McRuer, 1965) introduced in the previous chapter. The CO model hints at how humans adapt to different plants to elicit stable and effective control responses. Due to its simplicity, the model proposed aims to avoid common problems related to higher-complexity systems.

It was observed (McRuer, 1965) that when an external disturbance is introduced in the system, measured human operator responses were

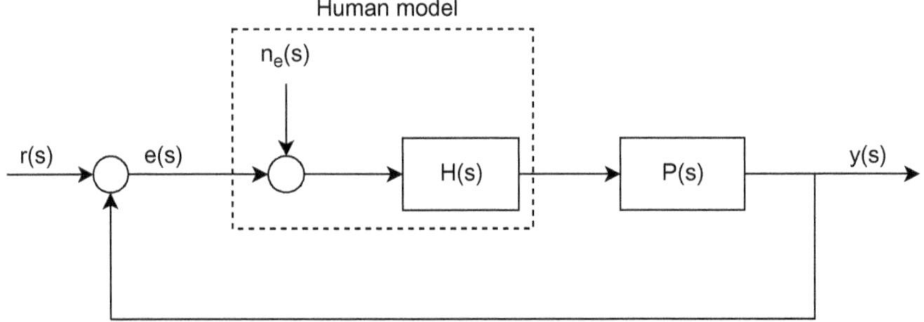

Fig. 3.1. Block scheme representing the structure of human–robot control system during a compensatory task, as described by the Crossover model. Here, $H(s)$ represents the human control dynamics, which is used along with the remnant noise $n_e(s)$ to build the human model. $P(s)$ represents the controlled element's dynamics (in our case, a robot).

different for different transfer functions of the controlled plant. However, the combined human–machine behavior is approximately the same for all the experiments.

As mentioned in the model introduction, the central theoretic hypothesis relying upon this approach is that the transfer function modulus bode plot has a -20 dB per decade slope around the CO frequency in the frequency domain. The remnant noise n_e (see Figure 3.1) models the non-stationary effects of the pilot behavior. A random process models such a linearly uncorrelated noise with the control input. Usually, the remnant is related to the error signal $e(t)$. It is, therefore, considered observation noise.

Given a controlled element, McRuer and Krendel (1974) proposed a more detailed precision model of human control action, which can be described in the Laplace domain by the following equation:

$$H(s) = K_p \frac{\tau_L s + 1}{\tau_I s + 1} \frac{1}{\tau_n s + 1} e^{-s\tau_e}. \tag{3.1}$$

Here, we have mainly four factors: a simple static gain element K_p, a lead-lag equalization term $\frac{\tau_L s + 1}{\tau_I s + 1}$, a time constant relative to the neuromuscular system τ_n and the so-called "effective time delay," being $e^{-s\tau_e}$.

Neuromuscular system dynamics have been extensively discussed by Magdaleno and Mc Ruer (1971), showing additional second-order characteristics typical of spindle and tendon organ ensembles. However, these effects are observable only in very high frequencies. Therefore, they can be neglected for most application scenarios, resulting in the first-order lag term in equation (3.1).

Moreover, the human operator sets the equalization term to achieve the -20 dB decade slope required by the CO theory. At the same time, the static gain is used to pilot the CO frequency value. According to McRuer, the sensitivity of the closed-loop stability on τ_L and τ_I should be very low, leaving the CO frequency and the effective time delay as the main parameters to consider to ensure stability and minimize the error.

The CO frequency mainly depends on human adaptation to the controlled element and, in a minor way, to the input bandwidth ω_i. Therefore, if we try to implement the same general concept expressed by equation (2.13) by including the controlled element dynamics into the precision model, we will have that the human–machine complex would be equivalent to

$$H(s)P(s) = K_{\mathrm{pc}} \frac{\tau_{\mathrm{LC}}s + 1}{\tau_{\mathrm{IC}}s + 1} \frac{1}{\tau_n s + 1} e^{-s\tau_e} \tag{3.2}$$

Here, τ_{LC} and τ_{IC} are the lead-lag equalization parameters considered after the initial training phase when the human learned the controlled machine dynamics and adapted its behavior to respect the CO hypothesis. The gain K_{pc} is now a general expression of the gain of the man-machine transfer function and is equivalent to the CO frequency.

The effective time delay τ_e results from transport delays of the sensory information to the central nervous system and high-frequency neuromuscular dynamics. McRuer and Jex (1967) gave the following empirical relationship:

$$\tau_e = \tau_0 - \Delta\tau(\omega_i). \tag{3.3}$$

Here, the first term τ_0 is the primary time delay when $\omega_i = 0$, depending on the controlled element Y_c. In contrast, the second term $\Delta\tau$ decreases the effective delay with the input bandwidth with a slope entirely independent of the controlled element. Such a relation is referred to as the scenario that led to McRuer's research activities in the aeronautic domain, where the human controller operates in high-frequency ranges. However, a constant behavior is expected if we focus on the primary delay term dependent on human adaptation to the controlled external dynamics.

Therefore, the reaction time delay should be constant for each human subject, with a small variability attributable to task and environmental variables. In particular, typical values of 0.5 s of the primary time delay τ_0

were obtained by McRuer and Krendel (1974) for second-order functions of the controlled element.

In this work, the precision human–machine model has been applied to a human–robot collaboration scenario to understand how to identify better human control characteristics concerning its adaptation to the controlled robot's dynamics. This adaptation process will be translated into properly tuning the model parameters, allowing the embedding of robot and human dynamics into a single transfer function, as done in the general CO model by McRuer (1965). Such modeling effort will mainly focus on reactive delay, an easily measurable parameter in which the discussed human behavior is expected to have an effect.

3.3   Design of Experiment

A simple manual guidance situation was considered, where humans and robots performed a typical point-to-point motion task with continuous physical interaction.

In this scenario, the reactive delay can be derived from the time displacement between the reference position signal and the actual measured position. The point-to-point task is the situation in which this displacement is primarily evident because it is constituted by a sequence of different target points that the human–robot complex has to reach.

Here, the reference signal to consider as the input $r(s)$ in Figure 3.1 is composed of a sum of steps being the series mentioned above of constant reference target points.

Figure 3.2 shows the position error derivative signal $\dot{e}(t)$ during a performed point-to-point task, with $e(t) = r(t) - p(t)$ being the position error, $r(t)$ the reference position, and $p(t)$ the actual position. As evident, $\dot{e}(t)$ is characterized by a series of local maxima, represented by the orange crosses. Each maximum corresponds to the time frame for the maximum variation in the reference position signal $r(t)$, i.e., corresponding to a target point variation, as will be better detailed in the following section. The green dots represent the following instant in which the error derivative becomes negative, caused by a variation in the actual position $p(t)$ in consequence of the human control action. Therefore, the difference between these two points can be considered a reliable experimental measure of the reactive time delay of the subject.

A Cartesian impedance controller was implemented on the robot side, depicted as the controlled element $P(s)$ in Figure 3.1. In the impedance

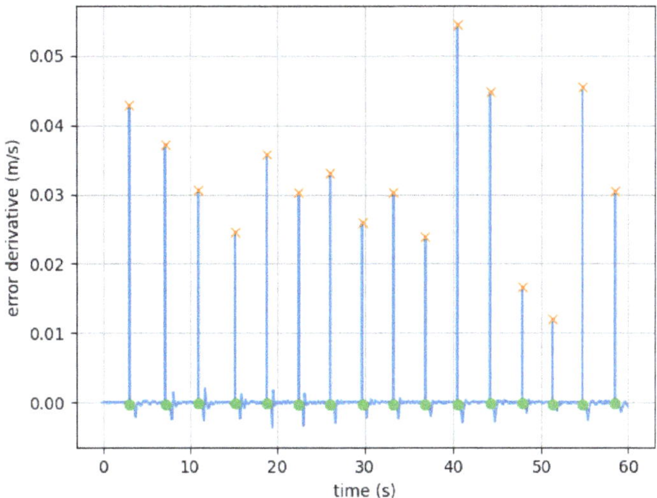

Fig. 3.2.   Time derivative of the position error signal in a performed experiment. Orange crosses represent local maxima, while green dots represent the first successive time instant where a sign change is observable.

control strategy, a virtual mass–spring–damper element is attached to the robot's end-effector's actual and reference positions. This way, when following a reference trajectory, any interaction with external forces causing deviations from the current path is managed by the system in a compliant way, depending on the parameters set on the virtual mass, spring, and damper.

To reproduce the best conditions to identify the human reactive delay, a simple compensation task was executed following the scheme depicted in Figure 3.3. The human operator had a continuous physical interaction with the robot. A constant low stiffness was set in the Cartesian impedance controller to ensure a compliant behavior of the robot.

In particular, the proposed experimental setup consisted of a UR5 robotic arm with a gripper in the end-effector, which the operator must hold. A virtual robot arm model was shown to the human operator on a screen using the Rviz software tool (Kam *et al.*, 2015). The reference target area, represented by a red square of 2 cm in such a virtual environment, was shown. Its position was randomly changed in the x–y plane (in the robot's base frame). The human operator was asked to reach the target point by manually guiding the robot gripper, holding it for $\Delta t = 3$ s, and bringing the robot back to the initial position. If the subject could not hold

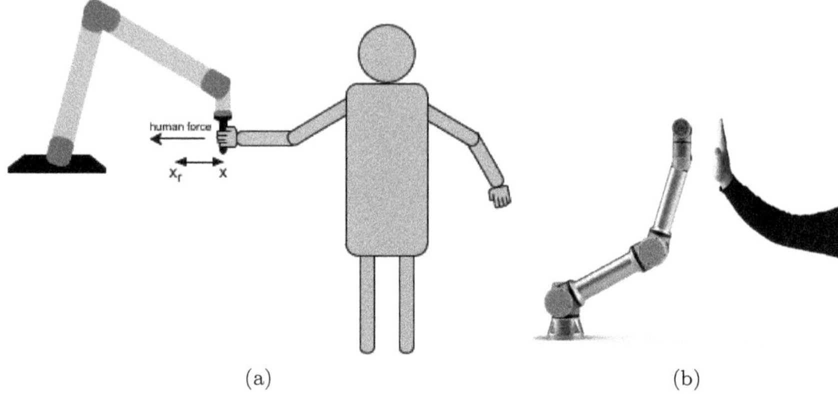

(a)   (b)

Fig. 3.3. (a) Scheme of the experimental setup: The disturbance signal causes a displacement from the initial reference position, and the human reacts by exerting a compensatory force toward the opposite direction. (b) Picture of the small UR5 robotic arm.

the robot end-effector within the reference target area for 3 s, the counter was reset, and the following virtual reference was not generated until the condition mentioned above was respected.

This operation was continuously repeated with different target points for 60 s. Each experiment was performed ten times by 10 human subjects with the same random target values. Before recording the experiments, a training phase was performed by executing the described operations until a high level of comfort with the environment was reached and confirmed by the subject.

3.4   Experimental Results

The recorded signals included many step motions since the human subject's reaction in both directions was considered when reaching the random target reference and returning to the initial position. The number of generated references depends on the subject's reaction delay and the precision when holding the robot into the target area. The total number of generated virtual references for each subject is indicated in the second row of Table 3.1, ranging from a minimum of 122 to a maximum of 151, considering all the 10 performed experimental trials.

To measure the reaction time delay, the displacement between each step of the reference position signal and the subsequent change of the actual position was considered. This interval corresponds, in other words, to the

Table 3.1. Average delay values for all the human subjects, considering the 10 performed experimental trials by each.

Human subject	1	2	3	4	5	6	7	8	9	10
Average delay	0.6234	0.4075	0.4777	0.4308	0.446	0.626	0.4494	0.5701	0.4086	0.6469
Generated ref. points	128	142	130	130	132	132	136	128	151	122

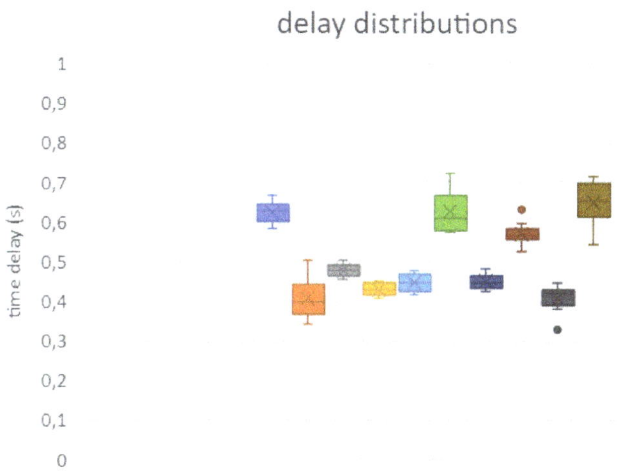

Fig. 3.4. Ranges of variation of mean reaction delay values considering the 10 performed experimental trials by each human subject.

difference between the time frame in which the system generates a new virtual reference point and the time frame in which the human subject starts to move the robot's end-effector.

Figure 3.4 shows a box and whiskers plot representing the range of variation of the measured reaction time delay for each human subject. Here, the vertical axis represents seconds. The mean value for each experiment of 60 s was computed to have a robust measure of such value. Then, the series of 10 mean values per subject were considered, and the resulting mean delays are summarized in Table 3.1.

As evident, only two outliers are present in all the measured values. Instead, most subjects display a narrow range distributed around the mean values, describing a constant delay around all the experiments. Moreover, such constant mean values significantly differ between one subject and another, confirming what was expected from the theoretical models.

In three cases, the ranges of values presented more consistent variations to the majority done by the other subjects. However, rather than being dependent on environmental or task variables, such behavior can be explained as an effect of the subject training and better adaptation to the external dynamics. This consideration is confirmed if looking at the values in detail.

Figure 3.5, for example, shows the series of measured data for subject 1 (Figure 3.5(a)) and subject 2 (Figure 3.5(b)), corresponding to the blue

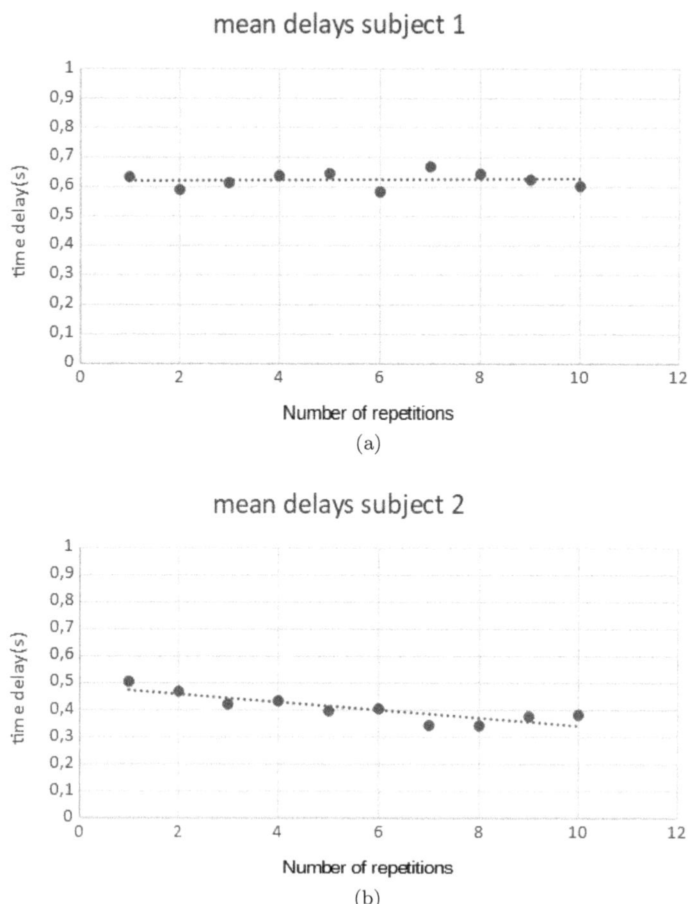

(a)

(b)

Fig. 3.5. Details of the measured delays for the first two subjects, boxes in blue and orange in Figure 3.4. The blue dots represent the mean value for each experimental trial, while the dotted line is the best linear approximation of the data trend.

and orange boxes in Figure 3.4. Here, the data trend confirms the above consideration, represented in its linear approximation by the dotted blue line. A narrow range corresponds to the expected constant behavior in the first case. At the same time, the wider range of the orange box in Figure 3.4 is caused by the descending trend observable in Figure 3.5(b), which testifies to the effect of the learning process of the external dynamics done by the subject.

The range of variation of the measured delays is between 0.4 and 0.6 s, which is compatible with the values obtained by the model. Further details will be given in the following section.

3.5 System Identification

All subjects' experimental data were compared with the simulated man–machine precision model results, as expressed in equation (3.2). The considered human model is constituted by five parameters: the gain element K_{pc}, time constants τ_{LC}, τ_{IC}, and τ_n, the effective time delay τ_e, each one having a different influence on the model simulated output. The static gain is responsible for the CO frequency value, and the time constants give the model response the typical second-order characteristics. In contrast, the effective time delay is the actual output parameter of interest to be compared with the measured values to check the model precision.

The same reference position signal, recorded from the performed experimental trials, was used as input for the simulated system, whose transfer function was implemented using the Python control system library. With $\hat{p}(t)$ denoting the simulated time forced response of the model, we will have a prediction error:

$$\hat{e}(t) = p(t) - \hat{p}(t). \tag{3.4}$$

Here, $p(t)$ is the actual position recorded from the experiments, representing the ground truth in this case. For each simulated response, the following prediction performance index (PPI) was defined:

$$\mathrm{PPI} = \int_0^{t_s} |\hat{e}(t)|. \tag{3.5}$$

Here, the integral of the error modulus was computed within the time window between 0 and t_s, corresponding to the maximum simulated time frame.

Table 3.2. Parameters of the human control model defined in equation (3.2) resulted in optimizing the prediction performance index in the performed simulations. All the values are expressed as an average between the ten performed experiments for all the human subjects. The last row reports the experimentally obtained effective delay τ_e to be compared with the simulated ones.

Human Subject	1	2	3	4	5	6	7	8	9	10
Crossover frequency	2.13	2.367	2.249	2.452	2.245	2.037	2.45	2.134	2.519	1.914
τ_n	1.46	1.34	1.384	1.151	1.427	1.594	1.29	1.548	1.252	1.905
τ_{LC}	0.86	0.912	0.865	0.908	0.877	0.871	0.89	0.818	0.947	0.741
τ_{IC}	0.996	0.994	0.991	0.991	0.991	0.996	0.99	0.995	0.996	1
Simulated delay τ_e	0.511	0.351	0.417	0.331	0.415	0.541	0.366	0.53	0.32	0.7
Measured delay τ_e	0.6234	0.4075	0.4777	0.4308	0.446	0.626	0.4494	0.5701	0.4086	0.6469

The five parameters were varied within reasonable intervals. Specifically, we used 0–10 Hz for the CO frequency, 0–2 s for the three time constants, and 0–1 s for the effective time delay τ_e. The simulation was run for each value of the five parameters, each run with one variable parameter and the other four fixed until the minimum prediction performance index was reached.

The results are summarized in Table 3.2, where the average values of the 10 experimental trials for each subject are reported for all the considered parameters. The CO frequency of the model is generally set around 2–2.5 Hz, confirming the low-frequency operational range of the model. The measured reaction delays obtained from the experiments are reported in the last row of the table. Comparing these values with the simulated reaction delays reported in the previous row, we can observe an error below 0.1 s. This result confirms the match between the identified model and the data.

Another reliable indicator of the model's accuracy is the comparison between the measured and simulated outputs, in our case, the actual position of the robot end-effector. Figure 3.6 compares the responses of the identified model (in green) and standard CO model (in orange), overlapped to the measured ground truth (in blue) recorded from the experiment, always considering the simulated parameters when the optimal PPI value has been reached. It is evident that, despite the good performance, the CO model's first-order characteristics fail to represent the first peak observable at each step of the real system response well.

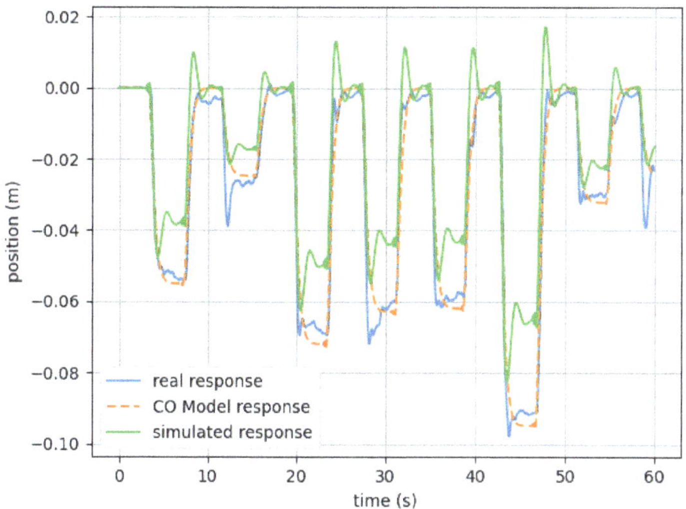

Fig. 3.6. Simulated position outputs considering the identified precision model (in green), standard CO model (dashed orange line), and measured position output (in blue). Despite the error in this example seeming to be averagely greater in the precision model, it is also evident how it succeeds in simulating more accurately the first-peak responses that the human generates after each step of the forcing function, which is the most important feature that is necessary for obtaining the human reaction delay.

On the other hand, the identified model response presents these second-order characteristics observable in the experimental data due to its lead-lag equalization term. Moreover, the model response is completely overlapped with the measured one in the transient phase of each step, which is crucial for obtaining a precise estimate of human reaction delay.

3.6 Final Considerations: Linear vs. Nonlinear Approaches

The linear approach of the proposed precision model was able to accurately describe human–robot physical interaction during a compensatory control task composed of different point-to-point motions to be performed under manual guidance by the human subject. Experimental trials involving ten healthy subjects of different ages were conducted to characterize human control delay according to the studied modeling approach.

The human effective reaction delay parameter was esteemed and extracted from raw post-process data to be reliable and invariant with task variables not dependent on the considered subject. The obtained results were within the expected values of 0.4–0.6 s, confirming the model's validity. Moreover, the system was identified by simulating a human–robot control model using the recorded experimental data. The model parameters were iteratively updated for each simulation to minimize a performance index defined for this purpose. The simulated results corresponded with the experimental data for the computed effective delay and the raw position output.

Despite the encouraging results, some aspects must be further investigated. For instance, for better practical applicability of this approach, the human control model should be improved for an online human intention estimation to allow the controlled robot to update its control behavior accordingly during the task. This last consideration was taken into account for the second nonlinear model, which will be proposed later in this work.

Until now, we have investigated and applied linear modeling techniques for characterizing human behavior when interacting with a controlled machine (in our case, a robot). Such techniques have proved to be useful in accurately representing the various physiological districts involved in the perception, interaction, and control processes. Moreover, in the previous chapter, we have shown how such a model can accurately simulate human response and experimentally identify the reaction delay when an external forcing function is applied to the system. However, linear models are not able to describe all the high-level processes that are involved in complex scenarios. To describe, in a deeper way, human adaptation to the plant when the complexity level of the task increases and to model also higher-level processes, such as decision-making, which are not correlated to traditional physiological subsystems, nonlinear models have to be necessarily investigated.

While linear models, such as McRuer's CO model, have historically dominated control theory due to their simplicity and analytical clarity, nonlinear models are becoming increasingly relevant in modern human–robot interaction applications. These approaches, including fuzzy logic and neural networks, capture the inherent complexity of human behavior more effectively. In the following, we compare the performance of both linear and nonlinear control models in key human–robot interaction scenarios.

As demonstrated, linear models excel in situations where the human's behavior follows predictable patterns, while nonlinear models are better suited for dynamic, uncertain environments, though they often require more data and tuning to perform optimally.

Chapter 4

Nonlinear Models

The investigations of human nonlinear dynamics when controlling a machine are so diversified that finding a common point between them is difficult. This chapter aims to give a structured overview of the existing techniques, focusing on the underlying physical and physiological human processes.

In this chapter, a human-centered review of the existing efforts will highlight the strengths and limitations of the presented techniques in their effort to model the intrinsic nonlinear dynamics in the human–machine system. Such nonlinearities will be referred to as spatial and temporal variables of functionals, which are identifiable within the human physiological control districts involved in sensing and information processing but also by deriving from the interaction with controlled element dynamics and/or the external environment (Salvendy, 1987). With respect to the existing review works related to each research field, this effort will help find a guiding line

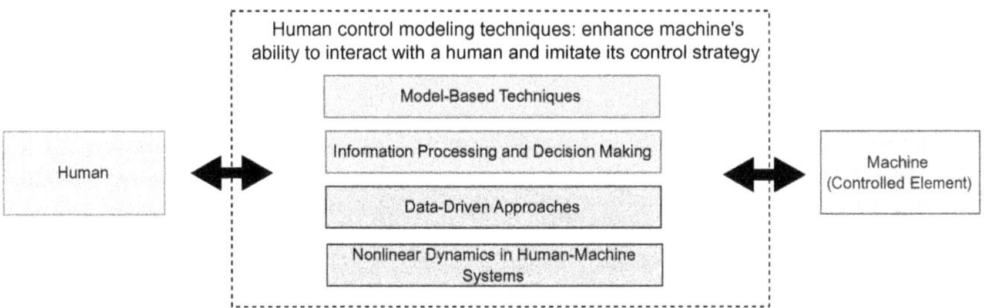

Fig. 4.1. Graphical abstract representing the structure of this chapter: How the study of nonlinear dynamics with different approaches and from different perspectives can be used to represent human behavior and increase the adaptability between the human subject and the controlled machine.

between more traditional control modeling techniques and modern learning algorithms.

The chapter is structured as follows: Section 4.1 discusses the dual-loop control model, giving an overview of the human controller when interacting with a controlled machine. Section 4.2 investigates the nonlinear muscular dynamics models. Then, in Section 4.3, the focus will be directed toward modeling techniques particularly useful for representing human information processing stages. Then, machine learning efforts to model the human decision-making process are investigated in Section 4.4. Lastly, in Section 4.5, practical examples of human–machine schemes in which the described techniques are applied to model complex nonlinear dynamics are considered.

4.1   Dual-Loop Control

In the later decades of the last century, Hess investigated human control strategies when interacting with a machine, resulting in his first "structural model" (Hess, 1980b). After the first linear version, Hess noted that often human operators' control strategies resulted in pulsive behavior, which was not linked to any feature of classical linear models. Hess (1979) linked the pulsive behavior to McRuer's quasi-linear model hypothesis in the frequency-domain context. The assumption was that when faced with a demanding task combined with the controlled element's high-order dynamics, the human operator avoids the computational effort and reduces the number of parameters by using a less computationally demanding nonlinear strategy rather than a linear one.

The above assumption was applied to the early version of Hess's model, resulting in the dual model depicted in Figure 4.2. Such a model resulted from an effort to link the hypothesis behind the crossover theory with the optimal control approach.

The described nonlinear factor results in the various switching elements in Figure 4.2. The first one (S1) allows selecting error or error rate tracking and is supposed to operate in unison with S2, which enables or disables the proprioceptive feedback loop. The physiological reason behind this is that after a triggering event, the pilot control strategy regresses to simple tracking behavior, where the error rate is controlled without the help of proprioceptive feedback. Right after the two described sensing channels, there is a time delay element due to the information processing occurring in the central nervous system, present right before the neuromuscular actuation and internal feedback stages.

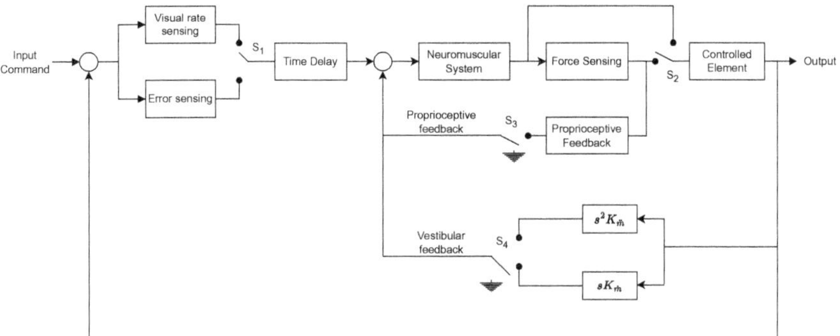

Fig. 4.2. Dual model of the human operator in a compensatory task.

Moreover, switch S3 allows modeling both displacement and force sensing inceptors. Ultimately, S4 allows using vestibular rate or acceleration inputs for control, with gain elements dependent on the perceived velocity $K_{\dot{m}}$ or acceleration $K_{\ddot{m}}$ (Hosman and Stassen, 1999). Ultimately, only neuromuscular and proprioceptive elements need parametrization, lowering the model complexity level. The neuromuscular block is often represented using second-order dynamics (McRuer *et al.*, 1968).

The neuromuscular force output is sensed and transformed into an estimation of the output rate of the controlled element using an internal model of its dynamics. This process is done by the proprioceptive system, which can be described in the Laplace domain using the following equation:

$$H_{\mathrm{ps}}(s) = \begin{cases} K_{\mathrm{ps}}(s+a) \\ K_{\mathrm{ps}} \\ \dfrac{K_{\mathrm{ps}}}{(s+a)}, \end{cases} \qquad (4.1)$$

where s is the Laplace variable in the complex plane and $a \in \mathbb{R}$. In other words, the proprioceptive system transfer function H_{ps} can be defined, depending on the controlled element dynamics, as a derivative term multiplied by a gain element K_{ps} (first case), through a simple proportional relationship (second case); or as integration (third case). If we indicate the controlled element transfer function as H_C, the proprioceptive system's dynamics would be chosen in order to satisfy the following relationship around the crossover frequency:

$$H_{\mathrm{ps}}(s) \propto sH_C(s). \qquad (4.2)$$

This concept well represents the operator's adaptability to external dynamics. This human's internal representation of machine dynamics expresses the hypothesis behind the crossover model and is equivalent to the Kalman estimator in the optimal control model (Baron *et al.*, 1970).

The last case of equation (4.1), in which the inner loop feedback signal is generated by integrating the force applied to the controlled element, is the one in which the effect of the pulsive control behavior on the time integrability of the human is more evident. In fact, the integration of a pulsive input signal can be approximated by

$$Y_{\mathrm{ps}} = \sum_{i=1}^{n} A_i \Delta T_i, \tag{4.3}$$

where A_i and ΔT_i are the equivalent calculated amplitude and time duration of the ith pulse, respectively, and Y_{ps} is the resulting proprioceptive output signal. The computational burden of such an operation when compared to integration over time is significantly lower. In order to represent the discussed pulsive control effect on the inner loop feedback in the most simple and realistic way, the following logic can be added before the neuromuscular system dynamics:

$$\frac{d\hat{q}}{dt} = 0 \quad \text{if} \quad \left|\frac{dq}{dt}\right| < \alpha$$

$$\hat{q} = \beta q \quad \text{if} \quad \left|\frac{dq}{dt}\right| \geq \alpha, \tag{4.4}$$

where q and $\hat{q}$ represent input and output variables, respectively, α and β are the only parameters that must be tuned to reproduce pulsive behavior. The dependence of the model on just two parameters allows it to avoid its over-parametrization and simplifies its adaptability to experimental data. This nonlinear element's action causes the output $\hat{q}$ to remain constant until a sufficiently rapid change in the input q occurs.

Pulsive control due to the "ease of integrability" principle, as hypothesized by Hess, found a physiological interpretation by Bachelder and Aponso (2022). In particular, while proportional and derivative control feedback can be actuated using direct sensing organs, such as muscle spindles and Golgi tendon organs, integral control does not have similar sensing input sources and requires higher-level cognition in the central nervous system (Hess, 1979). Consequently, when performing acceleration control, the human operator tends to generate a pulsive force rather than a continuous one to facilitate the integration process, the computational cost of the latter being much higher. Different explanations of the same phenomena are possible, for example, linked to energy-saving strategy when the required force peak value is low enough.

4.2 Neuromuscular Dynamics

The latter consideration suggests the importance of neuromuscular actuation mechanisms as a source of nonlinearity in the human controller. Several dynamical system modeling approaches of the neuromuscular system have been proposed in the literature, starting from simple state-space descriptions (Valero-Cuevas *et al.*, 2009). Neuromuscular dynamics are typically nonlinear; for instance, we consider the model of a human limb, and its characteristics can be described in state-space form as

$$\mathbf{x}_{t+1} = f(\mathbf{x}_t, t, \mathbf{u}) + \omega(t)$$

$$\mathbf{y}_{t+1} = h(\mathbf{x}_t, t) + \epsilon(t) \,, \tag{4.5}$$

where $\mathbf{x}$ is the state vector representing two angles and two angular velocities, $\mathbf{u}$ is the control input corresponding to the two applied joint torques, ω is the process noise, while ϵ is the observation noise. The general solution adopted in this nonlinear problem has been to linearize the nonlinear dynamics around a specific operating point in state space. The resulting linear time-varying dynamics can be used only in a small interval around the operating point; in the case of the above example, neglecting the noise terms would be equivalent to

$$\mathbf{x}_{t+1} = \mathbf{A}_t \mathbf{x}_t + \mathbf{B}_t \mathbf{u}_t$$

$$\mathbf{y}_{t+1} = \mathbf{H}_t \mathbf{x}_t \,. \tag{4.6}$$

Here, A is the state transition matrix and B the control transition matrix, while H represents the output measurement matrix.

Most control-theory-based neuromuscular modeling approaches aim to find the correct series of control inputs $u_1, \ldots, u_T$, corresponding to muscle forces and joint torques, making the system execute the desired trajectory in the time horizon $t = T$. Such a control system is an open loop; thus, if susceptible to disturbances, the controller would fail to reach the desired state, not sensing any state change. Moreover, the direct measurement of trajectories in state space can be problematic in high-dimensional systems, where part of the state may not be directly observable.

To overcome such shortcomings, optimal control approaches have been proposed (Bertsekas, 2012; Bryson and Ho, 2018), where the dynamical system is controlled by optimizing an objective function. According to optimal control theory, the controller can directly access output and state variables or estimate their values to implement an optimal control law to maximize the system's performance. A general mathematical expression of the objective function to be optimized to achieve this goal is

$$J(\mathbf{x}) = \min_{\mathbf{u}} \left(\phi(\mathbf{x_{t_N}}) + \int_{t_0}^{t_N} \left[q(\mathbf{x}) + \frac{1}{2} \mathbf{u}^T \mathbf{Ru} \right] dt \right). \tag{4.7}$$

The system variables are u, the control torques, forces, or neural commands, and x, often expressed as joint angles, velocities, or muscle activation. Moreover, ϕ is the cost term dependent on the state, describing how a given target was reached. At the same time, q is a state-dependent cost term considered over the whole time horizon t_N, and $u^T R u$ is the cost dependent on the control input (also considered over the time horizon t_N). The velocity value and the control effort used to perform a given trajectory can be good examples of the last two mentioned cost terms in a practical application.

Optimal control approaches for adapting classical linear techniques, such as linear quadratic Gaussian (LQG) regulator, have been proposed for nonlinear dynamics typical of muscles and multi-body limbs. Li and Todorov (2004) introduced an iterative linear quadratic regulator (ILQR) based on linearizing nonlinear muscular dynamics. An advantage of this approach is that it does not need any predefined target trajectory in the state space to work. The ILQR method was also extended by Todorov and Li (2005) for nonlinear stochastic systems characterized by state-dependent and control-dependent noise. Here, the ILQR technique permitted the description of the nonlinear relation between muscle force, fiber length, and contraction velocity. Further developments led to the use of extended Kalman filters (EKFs) in systems with additive noise in the sensory feedback loops (Li and Todorov, 2006, 2007).

4.3 Decision-Making and Information Processing

The discussed models helped describe the human–machine system dynamics in a control-theory fashion. The involved physiological districts, sensing, and actuation systems were put in relation, considering the human as an element of the control loop, and the nonlinear dynamics present in motion command actuation and feedback were put in evidence. However, to understand how human beings act as a controller when interacting with a controlled machine, a deep focus on the information processing stage is crucial to understand how its central nervous system integrates pieces of information to make decisions, learn, and generate commands.

4.3.1 *Fuzzy control models*

Processes such as human decision-making, inference, and judgment are challenging to characterize precisely. A modeling technique specifically meant to capture this concept is fuzzy control modeling. If we represent a human controller as a fuzzy subsystem, the core of its control model would be described by the fuzzy rules it will set. Specifically, fuzzy rules describe the human decision-making process starting from formulating a hypothesis and successively mapping the fuzzy set from an input to an output space (Zaychik *et al.*, 2006). Such a mapping process can be defined as the "fuzzification process," while the reverse transformation is called "defuzzification." The physiological equivalent of this process is when the neuromuscular system receives an abstract decision from the central nervous system and consequently emits a force to the controlled device/machine. Fuzzy logic control models have been used to represent various human control activities in many research works, achieving good results in overcoming the limitation of approaches relying on a strict categorical division, especially in classification problems. Miranda and Felipe (2015) used fuzzy logic classification to represent radiologists' reasoning and decision-making process when recognizing breast cancer types from the analysis of medical images. While Dovom *et al.* (2019) implemented a fuzzy architecture for malware detection and classification in IoT applications. In the aeronautic domain, fuzzy control is suitable for developing a mental model of the pilot during flight activity (Gestwa and Grigorie, 2011; Wohler *et al.*, 2014), primarily referring to a compensatory type of sub-tasks (George, 2008). The fuzzy logic control model was applied to study changes in simulated activity fidelity in aircraft control and dynamic multi-attribute decision making (DMADM) applications (Jiang *et al.*, 2014). Additionally, fuzzy control theory was used for the safety evaluation of landing operations considering aircraft (Su *et al.*, 2018; Yazdi *et al.*, 2017) and rotorcraft (Na and Lee, 2021). However, the computational efficiency of fuzzy control systems is limited in cases in which a vast number of rules is present (Stewart *et al.*, 2010). Moreover, it can be challenging to determine the rules when their number is high, and subjective model tuning will make its validation more challenging.

4.3.2 *Artificial neural networks*

For nonlinear dynamical system modeling, artificial neural networks (ANNs) have grown significantly in many research activities in the last few years due to their flexibility and ability to imitate human learning and decision-making. Moreover, when building a model from unstructured data, ANNs proved to be useful to build a reduced low-order model (Hinton and Salakhutdinov, 2006) and for their classification capability (Mori and Suzuki, 2009). ANNs are composed of a linear combination of fundamental units (i.e., neurons), which can provide a linear transformation from the input data $\mathbf{x}$ to output $\mathbf{y}$ through several intermediate hidden layers. Each ANN scheme can vary significantly if the input vector dimension is known. The user usually chooses the dimensionality of the hidden and output layers. The input–output relationship of a single-layer neural structure with m inputs (with m being a positive integer greater than 1) and a single output would be, in the linear case,

$$y = \sum_{i=1}^{m} \mathbf{x}_i \mathbf{w}_i + q \,, \tag{4.8}$$

where the variable $\mathbf{x}_i (i \in (1, 2, \ldots, m))$ represents the input signal of the model, y represents the output signal, $\mathbf{w}_i (i \in (1, 2, \ldots, m))$ is the weight of each input signal, and q is the threshold of the activation function f. Nonlinear activation functions can be used to represent a wider range of dynamics. The more general definition of an ANN constituted by M layers, providing a nonlinear mapping between input and output data, would be

$$\mathbf{y} = f_M(\mathbf{A_M}, \ldots, f_2(\mathbf{A_2}, f_1(\mathbf{A_1}, \mathbf{x})) \ldots) \,. \tag{4.9}$$

Here, the $A_M, \ldots, A_1$ matrices contain the weight coefficients w_i that map each variable from one layer to the next. The weights are chosen to fit the function

$$\mathrm{argmin}_{\mathbf{A_j}} (f_n(\mathbf{A_n}, \ldots f_2(\mathbf{A_2}, f_1(\mathbf{A_1}, \mathbf{x}))) + \lambda g(\mathbf{A_j})) \,. \tag{4.10}$$

Human behavior and information processing representation are based on the weights of neural networks. Such a modeling technique is advantageous in aeronautical applications, for example, when mapping pilot control in research works where extensive data to process are available (Jagacinski *et al.*, 2004). In the work by Tan *et al.* (2003), ANN and quasi-linear approaches are confronted in a two-axis tracking task, verifying neural network accuracy in describing nonlinear pilot behavior in aircraft control. Cheng *et al.* (2021) used an adaptive neural network controller by combining the trained network and a proportional-integral controller in an attempt to find a model-based method for control determination of unknown dynamics.

4.3.3 *Neuro-fuzzy systems*

Generally, neuro-fuzzy systems can be defined as all the modeling techniques involving ANNs and fuzzy logic. These techniques can be categorized into the following three classes, depending on the combination of the two elements (Viharos and Kis, 2015):

- cooperative neuro-fuzzy systems,
- concurrent neuro-fuzzy systems,
- hybrid neuro-fuzzy systems.

In a cooperative system, the neural component is only present in an initial phase and determines the blocks composing the subsequent fuzzy system using training data. After this stage, only the fuzzy system will be executed.

In concurrent systems, on the other hand, the neural and the fuzzy components work simultaneously. This means that the information is pre-processed by one of the two components and then given as input to the other.

The most promising and utilized models belong to the hybrid systems category. A hybrid neuro-fuzzy system can be imagined as a fuzzy system in which parameters, such as fuzzy sets and fuzzy rules, are determined using a learning algorithm inspired by the neural network theory. Such a neuro-fuzzy system can be entirely created starting from measured input-output data without the *a priori* knowledge needed to develop fuzzy rules with the traditional approach.

An example of a commonly used model of this type is the adaptive-network-based fuzzy inference system (ANFIS), which was proposed for the first time in 1993 by Jang. Its structure is composed of five layers. The first hidden layer maps the input variable relative to each membership function. The output layer calculates the global output as the summation of all the signals coming in the input. In particular, input membership function parameters are determined using backpropagation learning algorithms, and the least mean square method is used to determine the consequent parameters. The first advantage is to show both characteristics of neural networks and fuzzy logic, comprising if-then statements more suitable for human-like decision-making logic. In addition, its structure is not a black box, as in the case of neural networks, and therefore can be more easily debugged and improved. Moreover, it has smaller parameters to be determined to provide faster training without loss of generality (Ertugrul, 2008). Such a model recently found diverse domains of application aside from human–machine interaction, such as electric distribution systems (Abbas *et al.*, 2022; Ying and Pan, 2008), speech recognition (Avci and Akpolat, 2006), and economics (Wei *et al.*, 2011). Kordnoori *et al.* (2022) used the ANFIS model for human fall detection in comparison with other neuro-fuzzy techniques, such as the local linear model trees (LOLIMOT) model (Murray-Smith, 1994).

In LOLIMOT models, each neuron is a local linear model (LLM), and an associated validity function determines the region of validity of the LLM. The normalized validity functions form a partition of unity for any model input z are

$$\sum_{i=1}^{M} \phi_i(\mathbf{z}) = 1. \tag{4.11}$$

While the output of each LLM is calculated as follows:

$$\hat{\mathbf{y}} = \sum_{i=1}^{M} (\omega_{i,0} + \omega_{i,1} x_1 + \cdots + \omega_{i,nx} x_{nx}) \phi_i(\mathbf{z}), \tag{4.12}$$

where $\mathbf{x} = [x_1, x_2, \ldots, x_{nx}]^T$. Here, the LLMs depend on $\mathbf{x}$, while the validity functions depend on $\mathbf{z}$ and are typically chosen as normalized Gaussian. The overall LOLIMOT network output is computed as a weighted sum of the LLMs outputs, where $\phi_i(0)$ can be interpreted as the operating point-dependent weighting factors. The network interpolation between different LLMs is performed with the validity functions, where weights $w_{i,j}$ are linear network parameters. Again, LOLIMOT models were used in various domains of application, such as transportation (Aghabayk *et al.*, 2013), medicine (Salmanpour *et al.*, 2020), complex systems (Iranmanesh *et al.*, 2016), and identification of time-variant nonlinear dynamics (Nelles, 1996).

A dual fuzzy neural networks (DFNNs) model composed of two equal neural networks has been used to simulate the physical nervous system by

Xu (2012a). The advantage of DFNNs is related to their close similitude to functioning and flexibility typical of humans. As in the case of ANNs, DFNNs can choose a suitable nonlinear mapping of input/output features through an iterative learning phase, in which neuron weights are updated. Such a model was implemented to simulate the relationship between the control signal and human-perceived input (Xu, 2012b). Its performance was evaluated to provide insight into the pilot's decision-making process (Xu *et al.*, 2012). Moreover, Zhu *et al.* (2012) proposed a risk evaluation procedure founded on ANNs with the fuzzy control approach.

Even though in the neuro-fuzzy model, neural networks and fuzzy logic are integrated, its drawback is to increase the computation and tuning time potentially. Besides, the experimental validation of the obtained model parameters may be tricky.

4.4 Data-Driven Approaches

Data-driven approaches have attracted more and more attention in recent years in various application scenarios in nonlinear dynamical system modeling and identification (LeCun *et al.*, 2015). In human–machine interaction, learning processes starting from unstructured data using different types of ANNs have been used for their processing classification by imitating human learning capability and decision-making, often combined with other learning algorithms, as we will discuss later in this section. A widely used type of network is the convolutional neural network (CNN), traditionally used for capturing spatial relations in data, valid for applications with robust image processing, which are very common in human–robot collaboration (Nicola *et al.*, 2022) or autonomous system navigation (da Silva *et al.*, 2022).

However, for the study of nonlinear dynamics introduced by the human into the system during its control activity, aspects such as its temporal delay (Scibilia *et al.*, 2022b) or temporal relations in general within the given data series, might be more relevant. Recurrent neural networks (RNNs) are the primary ANNs suitable for processing time series and other sequential data types. RNNs can extract a sequence's contextual information by defining the mutual dependencies between various time stamps. As shown from the scheme represented in Figure 4.3, a standard RNN is composed of numerous successive recurrent layers and has a lot of feedforward and feedback connections in the time direction, allowing it to sequentially model its layers to map a sequence with other sequences. This makes it a good choice for dynamic system identification and control.

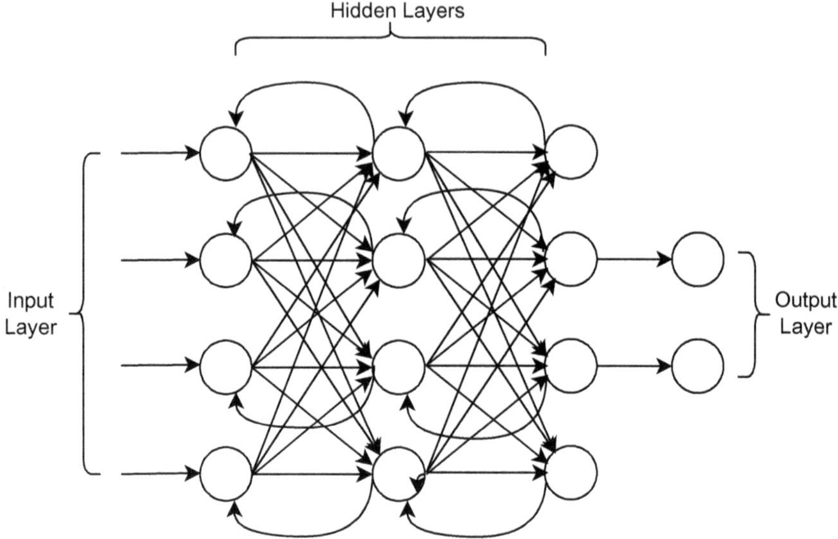

Fig. 4.3.   General structure of a recurrent neural network and its internal feedbacks.

Concerning its structure, an RNN can be defined as an extension of feedforward ANN with internal loops in hidden layers. The activation of the state of a recurrent hidden layer at each time instant is dependent on that of the previous one. At a given time frame, each non-input unit computes the current activation as the nonlinear function of the weighted sum of all the activations of every connected unit (Akash *et al.*, 2021). They have been successfully applied in natural language processing (NLP), image captioning, speech recognition, and other fields. Li *et al.* (2005) investigated the approximation capability of continuous-time RNNs to the time-variant dynamical systems. They proved that such network performance for approximating any finite time trajectory of a time-variant system was high. However, despite its suitability to model temporal variations present in the input, depending only on the current information and the previous output, a standard RNN may encounter difficulties when it comes to capturing long-term dependencies of time sequences.

To overcome this limit, a popular type of RNN which was proposed in many studies is the long short-term memory (LSTM) network. An LSTM network is a modified RNN, mainly designed to improve its ability to capture long-term relationships by avoiding premature gradient disappearance in error backpropagation algorithms through time.

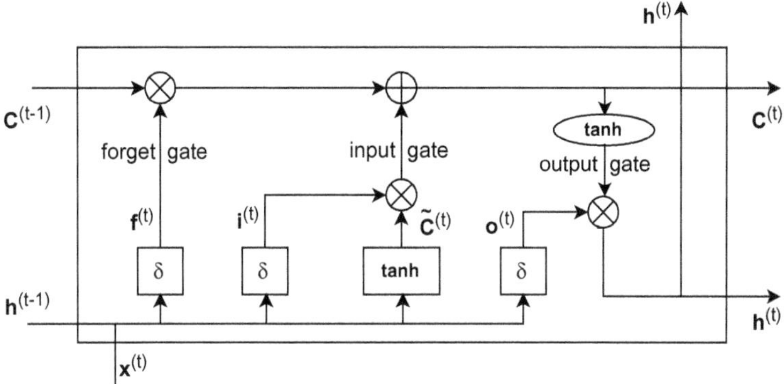

Fig. 4.4.   Internal structure of an LSTM unit (Chen *et al.*, 2022).

LSTM is composed of a combination of units, representing internal structure in Figure 4.4. Each unit simultaneously receives an input vector $\mathbf{x}^{(t)}$ and the state of the hidden layer in the previous time instant $\mathbf{h}^{(t-1)}$ and updates, as output information, the cell state $\mathbf{C}^{(t)}$ and the current state of the hidden layer $\mathbf{h}^{(t)}$. This operation is done through three embedded layers in each LSTM unit: the input, output, and forget gates. The three gates have different roles and work in coordination: The forget gate $\mathbf{f}^{(t)}$ determines the probability that certain information has to be canceled from the cell state vector; the input gate $\mathbf{i}^{(t)}$ identifies the new information to be stored, while the output gate $\mathbf{o}^{(t)}$ controls the output of the current hidden state $\mathbf{h}^{(t)}$. Translated into mathematical expressions, the LSTM unit operations are the following:

$$\mathbf{f}^{(t)} = \sigma(\mathbf{W_f}\mathbf{h}^{(t-1)} + \mathbf{U_f}\mathbf{x}^{(t)} + \mathbf{b_f})$$

$$\mathbf{i}^{(t)} = \sigma(\mathbf{W_i}\mathbf{h}^{(t-1)} + \mathbf{U_i}\mathbf{x}^{(t)} + \mathbf{b_i})$$

$$\tilde{\mathbf{C}}^{(t)} = \tanh(\mathbf{W_C}\mathbf{h}^{(t-1)} + \mathbf{U_C}\mathbf{x}^{(t)} + \mathbf{b_C})$$

$$\mathbf{C}^{(t)} = \mathbf{C}^{(t-1)} \odot \mathbf{f}^{(t)} + \mathbf{i}^{(t)} \odot \tilde{\mathbf{C}}^{(t)}$$

$$\mathbf{o}^{(t)} = \sigma(\mathbf{W_o}\mathbf{h}^{(t-1)} + \mathbf{U_o}\mathbf{x}^{(t)} + \mathbf{b_o})$$

$$\mathbf{h}^{(t)} = \mathbf{o}^{(t)} \odot \tanh(\mathbf{C}^{(t)}). \tag{4.13}$$

where $\mathbf{W}$, $\mathbf{U}$, and $\mathbf{b}$ represent, respectively, the recurrent matrix, input weight matrix, and bias vector, σ and *tanh* are the sigmoid and hyperbolic tangent functions, and $\odot$ represents the element-wise Hadamard product.

Yeo *et al.* (2019) implemented LSTM networks to build a simulation model of noisy nonlinear dynamical systems using experimental data. Their goal was to identify the best fit of the probability density function of a given stochastic process and to represent the underlying nonlinear dynamics. Chen *et al.* (2022) used LSTM networks to learn the characteristics of strongly nonlinear external dynamics of Van der Pol and Lorenz systems.

As mentioned, neural networks used for unstructured learning have increased their potential by combining them with other learning algorithms. The most promising technology in this sense is reinforcement learning (RL).

Unlike supervised and unsupervised learning, RL has emerged as the third kind of machine learning paradigm. Using computational RL algorithms allowed us to quantitatively describe several previously abstract concepts in neuroscience, cognitive science, and behavioral science (Matsuo *et al.*, 2022).

As detailed by Sutton and Barto (2018), RL can rely on Markov decision processes as a learning framework in which a learning agent interacts with an external environment and perceives its state, choosing its actions to maximize a numerical reward function. The reward function is a simple numerical value for each time stamp, which can increase or decrease by one unit in the future due to the agent's actions. Therefore, the goal to maximize the reward function can be translated into maximizing the expected value of the cumulative sum of the scalar reward signal. Being defined from external information acquired from the environment through sensory inputs (in the case of a human operator), the goal to achieve is always defined outside the learning agent. In the case of a human being, that means that the learning agent can be defined as only the subsystem deputed to process the external inputs to define a control strategy (i.e., the central nervous system). The sensory subsystems can be considered part of the environment. In real-world complex situations in which humans are confronted with a challenging task, their duty is to derive efficient representations of the environment from high-dimensional sensory inputs and use them to generalize past experience and be able to use it in new situations (Mnih *et al.*, 2015). If we consider an episodic task in which the agent–environment interaction can be decomposed into sub-sequences of repeated interactions, there is also a final time step, T. In this case, for a given timestamp t, the reward function to maximize is

$$G_t = \sum_{k=0}^{T-t-1} \gamma^k R_{t+k+1}, \qquad (4.14)$$

where γ, with $0 \le \gamma \le 1$, is the discount rate. This parameter determines the present value of future rewards. When γ is close to zero, the weight of immediate rewards is higher and mostly taken into account by the agent; as it approaches 1, the goal takes future reward values more strongly weighted. If we have a continuous interaction in which there are neither definable intermediate steps nor a known final time frame, the above equation can be rewritten with $T = \infty$.

RL algorithms were extensively used in many research works relating to humans interacting with a machine, with many reward functions designed and more suitable for the different application scenarios. Zhang *et al.* (2022b) used a deep deterministic policy gradient (DDPG) RL algorithm is used to estimate human intentions in a human–robot interaction framework using EMG sensory inputs. At the same time, Li *et al.* (2022) integrated RL into the robot motion planning in a multi-robot collaborative manufacturing plant to implement human-in-the-loop control in teleoperated robots through augmented reality and digital twin techniques.

In the transport field, Wang *et al.* (2022b) adopted microscopic traffic simulation and RL to implement the lane-changing strategy in connected and automated vehicles (CAVs). RL has been successfully used with model-based techniques for systems identification by Self *et al.* (2022). This was done to estimate the reward function from online data by acquiring and processing linear and nonlinear external dynamics. Mu *et al.* (2020) used a RL algorithm for partially non-modeled nonlinear systems, coupled with two neural networks, to implement an event-triggering dynamic strategy. In robotics, deep RL can be used for motion planning in cooperative applications with a human subject, learning how the human interacts with a specific environment and adaptively computing the best way to interact with him (Nicola and Ghidoni, 2021).

4.5 Nonlinear Dynamics in Human–Machine Systems: The Case Study of Autonomous Driving

The discussed modeling techniques and data-driven approaches have been successfully used for describing nonlinear dynamics in many application domains where humans interact with a controlled element.

Transport systems, for example, are a particularly relevant field of application for nonlinear dynamics modeling in human–machine interaction for what concerns the nonlinear dynamics deriving human decision-making,

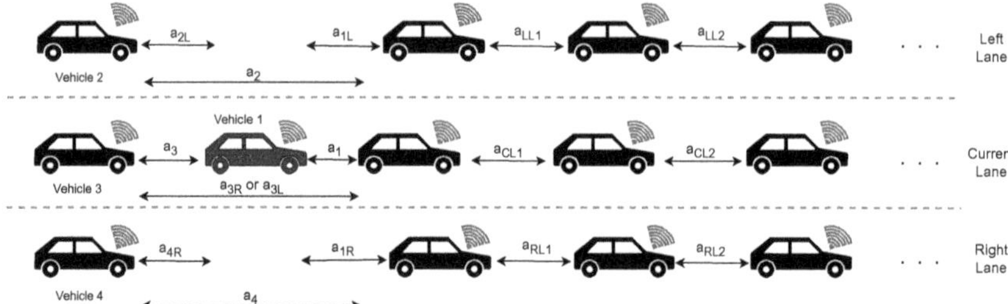

Fig. 4.5.   Connected automated vehicles represented in a lane-change scheme.

from the nonlinear nature of the controlled element and/or the system, and from human body physical coupling with the controlled system.

For what concerns the first point, CAV development has gained more and more attention from companies and research centers in the last few years. In studies dealing with automated lane changing, machine learning techniques were extensively used for human decision-making modeling and their use in automatic control strategies.

Let us consider the situation described by Figure 4.5, in which vehicle 1 (V1) has to choose a lane-change strategy and is followed by vehicles 2, 3, and 4. If we discretize the CAV travel as a series of time steps t, and S_t is the state of the external environment at each step t, we would have that

$$S_t = \{\tilde{a}_{1L}, \tilde{a}_{1R}, \tilde{a}_{2L}, \tilde{a}_{4R}, \tilde{a}_3, \tilde{\nu}_L, \tilde{\nu}_R, \tilde{\sigma}_L, \tilde{\sigma}_R\}(t) \,. \tag{4.15}$$

Here, $\tilde{a}$ represents the acceleration difference of the considered vehicles (in the subscripts, numbers represent the vehicle and letters the lane change direction) after V1 lane change; $\tilde{\nu}$ represents the mean acceleration difference between the central and the left or right lanes; $\tilde{\sigma}$ represents the difference between the standard deviations of the acceleration differences. From a learning agent perspective, these acceleration differences represent the gain obtained after a lane change. Therefore, the reward function could be formulated as follows:

$$R_t = a_1^{t+1} \,. \tag{4.16}$$

The subscript number stands for vehicle 1, and $t + 1$ represents two consecutive simulation time stamps.

Many research efforts on this topic used simulation environments such as Matlab toolboxes (Du *et al.*, 2014; Nilsson *et al.*, 2016) to represent vehicles' behavior, or robotic toolkits using partially observable Markov decision processes (POMDPs), such as by Liu *et al.* (2015).

A CAV does not rely on any external supervisor but must autonomously learn with a trial-and-error approach to decide when to make a lane change and how to execute it. One of the most challenging aspects is that the vehicle must evaluate the long-term benefits of such an action and become farsighted in its strategy to maximize the travel's efficiency. For this challenge, RL seems to be the preferential approach (as observed from its formulation described in the previous section). For instance, in a high-fidelity simulation environment, Ye *et al.* (2019) used a deep RL training program for car following. Wang *et al.* (2022b) also used RL in a microscopic traffic simulation environment (Anya *et al.*, 2014) calibrated using actual highway data. Li *et al.* (2019b) used an evolutionary learning approach for lane change tested in a highway simulation environment. The optimization problem objective is to maximize the velocity while minimizing the disruption to the following vehicle; if it is impossible to reach this goal in the current lane, a change-lane decision is taken.

In this case, the reward $r_{i,t}$ depends on the difference between desired velocity v_d and actual velocity $v_{i,t}$ of the controlled vehicle and the acceleration of the following one ($a_{k,t}$):

$$r_{i,t} = -|v_{i,t} - v_d| + a_{k,t}. \tag{4.17}$$

If the velocity difference overcomes a certain threshold, the lane is changed. The decision-making process has indeed a purely nonlinear behavior, also reflected by the resultant reward values during the training of the system, as shown in Figure 4.6.

However, the lane change has not a time-driven structure but an event-driven one, described as a discrete dynamic process, which can be well represented as a Markov decision process. In the study by Ulbrich and Maurer (2015), POMDPs were also used for an automatic lane change in long-distance road experimental trials using automated vehicles. Here, the decision-making process is modeled, referring not only to the controlled vehicle but also to the surrounding environment, inspired by the consideration that human drivers change their behavior when interacting. Reaction modeling is performed by measuring the temporal evolution of the vehicle state, including in it also a reaction and a deviation parameter as well.

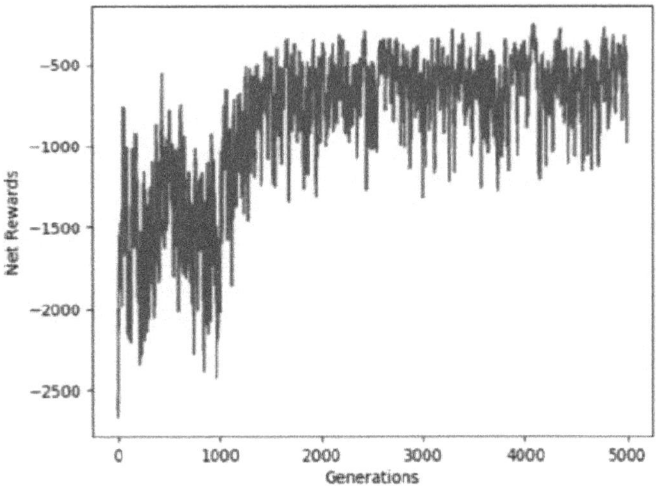

Fig. 4.6.   Reward values during the system training in evolutionary learning adopted by Li *et al.* (2019b).

Figure 4.7 presents the differences between the traditional hidden goal method, which applies only to specific regions of interest, and the reactive method, which models the group of vehicles in general and their deviation.

A further aspect of human decision-making in a lane-changing application is related to risk propensity. Jiang *et al.* (2022) proposed a decision model that considered the driver's perception, reasoning, and emotions. Risk propensity considers two mental processes: regret biasing and probability weighting, corresponding to the emotional aspect and cognitive reasoning. Both functions' nonlinearity increases proportionally to the emotional bias and cognitive weighting. The proposed model was tested with a dataset from a naturalistic driving database. Figure 4.8 presents the obtained fitted functions without considering regret biasing and probability weighting (purple line) and with the two terms (green dashed line). Figure 4.8(a) presents the regret q-functions, which resulted in being linear, indicating the regret influence is not evident in all cases. On the other hand, Figure 4.8(b) shows a w-function divided into three intervals; in two of them, the function overweights the objective probability (dot-dashed line), indicating a general optimism and bent to take risks.

As mentioned, aside from human decision-making representation, the source of nonlinearity in the human–machine complex may be related to the dynamics of the controlled element. If the human subject continuously

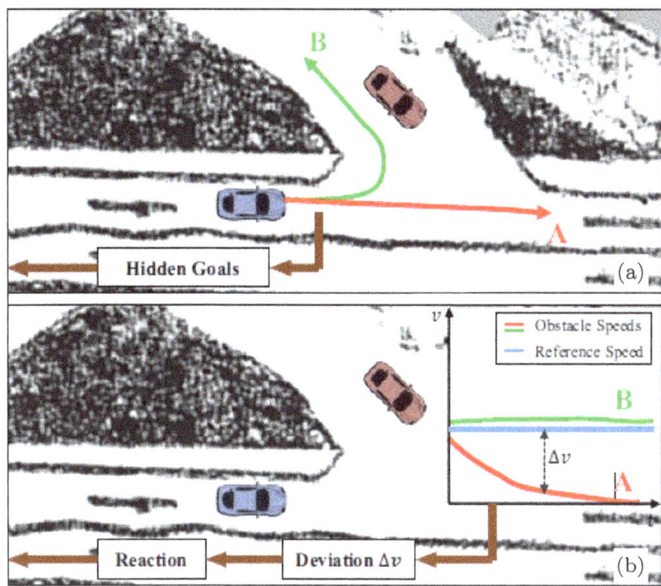

Fig. 4.7. Motion intention estimation as explained by Ulbrich and Maurer (2015), using hidden goal (a) and reaction (b) methods.

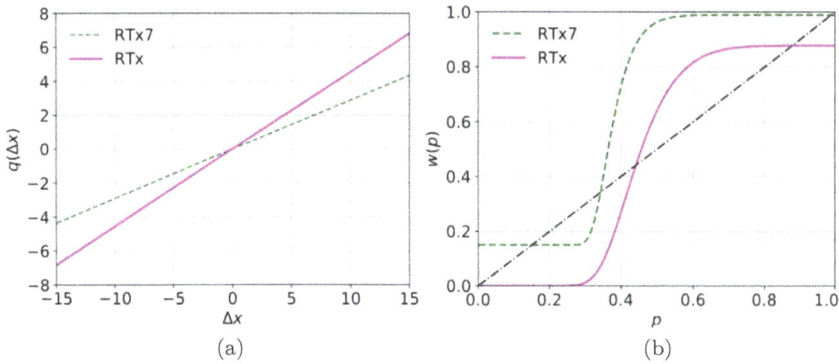

Fig. 4.8. Modeling of regret biasing (a) and probability weighting (b) at a cognitive level as studied by Jiang *et al.* (2022).

controls such devices, this will raise an essential challenge concerning system modeling and control.

In cooperative teleoperated robotic systems, for instance, many nonlinear control approaches have been developed in order to deal with non-passive (and therefore unstable Lawrence, 1993) factors such as the uncertainty of the environment, the presence of variable communication delays, kinematics, and dynamics parametric uncertainty. Such kinds of systems have found vast applications in healthcare (Murphy and Alambeigi, 2023), space (Gao *et al.*, 2023), exploration in dangerous environments (Szczurek *et al.*, 2022), and disaster scenarios (Kashiri *et al.*, 2019). Even if linear control approaches have been successfully developed for robust stability achievement in the presence of uncertain system dynamics, nonlinear controllers proved to guarantee good stability and performance through the exploitation of special properties of nonlinear rigid body dynamics of master and slave manipulators (Sirouspour, 2005).

Rashidinejad *et al.* (2015) performed nonlinear bilateral control of a teleoperation system with a flexible-link slave manipulator by designing a robust tip position tracking controller for the slave manipulator. The desired trajectory is determined based on the master's position signal and a force controller for the master robot, which should track the environmental force exerted on the slave manipulator. Moreover, Sirouspour and Setoodeh (2005) proposed a control strategy able to establish position–position kinematic correspondence between master and slave by incorporating in the adaptive controller the models of operators, controlled robots, tools, and environment, as well as their parametric uncertainty. Further approaches, such as that by Ajoudani *et al.* (2012), extended this concept by mapping the human arm stiffness references in a bilateral teleoperation framework, building a "teleimpedance control," later extended with a semi-autonomous contact detection strategy by Scibilia *et al.* (2018). Moreover, another challenging aspect of bilateral teleoperation systems control is related to the presence of communication time delays, which may cause the system to degrade its performance and even result in unstable behavior. The time delays should therefore be considered in the design stage of the controller. Wang *et al.* (2017) pursued this problem by considering adaptive neural synchronization control of bilateral teleoperation systems with backlash-like hysteresis, one of the most important nonlinearities in robots, while Yang *et al.* (2013) proposed a finite-time synchronization control method based on fuzzy approximation of system uncertainties.

Another example of highly nonlinear controlled systems interacting with an unknown external environment consists of multirotor remote control. Multirotor applications were carried out in several research activities, with practical applications like surveillance, photography, video-making, grasping or motion of an object, or military (Lee and Kim, 2017). The equation of motion of a multirotor with a mass m and inertia tensor $\mathbf{J}$ can be written as

$$m\ddot{\mathbf{x}} = -mg\mathbf{e_3} + f\mathbf{R}\mathbf{e_3}$$

$$\mathbf{R} = \mathbf{R}\hat{\boldsymbol{\Omega}}$$

$$\mathbf{J}\dot{\boldsymbol{\Omega}} = -\boldsymbol{\Omega} \times \mathbf{J}\boldsymbol{\Omega} + \boldsymbol{\tau}, \tag{4.18}$$

where $\mathbf{f}$ and $\boldsymbol{\tau}$ are the force and torque inputs, respectively, $\mathbf{x}$ is the multirotor position with respect to the inertial frame, $\boldsymbol{\Omega} = [p_B, q_b, r_B]^T$ is the angular velocity vector in the body frame, $\mathbf{g}$ is the gravity force, $\mathbf{e_3} = [0, 0, 1]^T$, and $\mathbf{R}$ is the transformation matrix from the body to an inertial frame.

The hat superscript indicates the transformation map between a vector in $\mathbb{R}^3$ and a 3×3 matrix. The translational dynamics will be

$$\ddot{\mathbf{x}} = \frac{\mathbf{f}}{m}u_x, \quad \ddot{\mathbf{y}} = \frac{\mathbf{f}}{m}u_y, \quad \ddot{\mathbf{z}} = \frac{\mathbf{f}}{m}u_z - \mathbf{g}, \tag{4.19}$$

where ψ, θ, and ϕ are yaw, pitch, and roll, respectively, and $u_x = (\cos_\psi \sin_\theta \cos_\phi + \sin_\psi sin_\phi)$, $u_y = (\sin_\psi \sin_\theta \cos_\phi - \cos_\psi \sin_\phi)$, and $u_z = (\cos_\theta \cos_\phi)$.

If $\boldsymbol{\nu} = [\psi, \theta, \phi]^T$, $\mathbf{T}_\nu$ is the Jacobian to convert $\boldsymbol{\Omega}$ to $\dot{\boldsymbol{\nu}}$, and $\mathbf{J}_\nu = \mathbf{T}_\nu{}^T \mathbf{J}\mathbf{T}_\nu$ the rotational inertia tensor, the rotational dynamics can be expressed with a Lagrangian formulation as

$$\mathbf{J}_\nu \ddot{\boldsymbol{\nu}} + \mathbf{C}(\boldsymbol{\nu}, \dot{\boldsymbol{\nu}})\dot{\boldsymbol{\nu}} = \boldsymbol{\tau}, \tag{4.20}$$

where C is the Coriolis matrix. If the roll and pitch angles are small, (4.20) can be simplified into

$$\ddot{\phi} = \mathbf{I_{xyz}}\dot{\psi}\dot{\phi} + \frac{\tau_\phi}{\mathbf{I_{xx}}}, \quad \ddot{\theta} = \mathbf{I_{xyz}}\dot{\phi}\dot{\psi} + \frac{\tau_\theta}{\mathbf{I_{yy}}}, \quad \ddot{\psi} = \mathbf{I_{xyz}}\dot{\phi}\dot{\theta} + \frac{\tau_\psi}{\mathbf{I_{zz}}}, \tag{4.21}$$

where $\mathbf{I_{xyz}} = \frac{(\mathbf{I_{xx}} - \mathbf{I_{yy}})}{\mathbf{I_{zz}}}$.

Trajectory tracking control for such systems is not accessible due to their nonlinearity, under-actuation, and highly coupled states. Although simple linear controllers such as PID or LQR have been successfully proposed in the past (Bouabdallah *et al.*, 2004; Cowling *et al.*, 2010; Dydek *et al.*, 2012; Hoffmann *et al.*, 2008) for a limited number of non-agile movements, controllers using feedback linearization (FL), backstepping, or geometric control techniques are more suitable to handle the nonlinearity of the system. Various types of FL techniques were used for multirotors, such as input–output and state-space linearizations (Henson and Seborg, 1997), have been used for finding the rotor's dynamics linear approximation. For instance, such a linear relationship can be obtained by differentiating (4.19) and ψ in equation (4.21) until the control input terms are explicitly expressed. Controllers designed with this kind of procedure, however, will present high-order derivative terms, which might cause the controller to be less robust with regard to sensor noise. To derive the input of the position controller, for example, (4.19) is differentiated until the fourth order until the control input term shows up. Attitude controllers in roll, pitch, and yaw angles can be designed similarly. These high-order derivative terms cause performance degradation of the controller as a consequence of disturbances such as the model's uncertainty. Despite this, FL showed promising results for finding the rotor's dynamics linear approximation. Lee *et al.* (2009) compared FL performance with an adaptive sliding mode control technique, while Mokhtari *et al.* (2006) combined an FL controller with a Luenberger observer.

However, the rotorcraft nonlinearity cannot be eliminated if a modeling error is present in FL. Thus its stability is not guaranteed. Therefore, backstepping control strategies with sliding mode techniques have been increasingly used to overcome these problems associated with sliding mode techniques by Castillo *et al.* (2006), Bouabdallah and Siegwart (2005), and Dikmen *et al.* (2009). A backstepping controller with sliding mode techniques is used by Castillo *et al.* (2006), Bouabdallah and Siegwart (2005), and Adigbli *et al.* (2007), based on the nonlinear translational (4.18) and the rotational simplified (4.21). When roll and pitch angles were high, such as in the works of Madani and Benallegue (2006) and Das *et al.* (2009), the Lagrangian formulation was preferred, even at a higher computational cost. Xian *et al.* (2019) proposed a different approach in which an energy-based passivity controller controlled a quadrotor with a suspended payload. Also, neural networks were used in multiagent trajectory tracking applications, such as that reported by Hwang and Abebe (2021), where an online RNN-based controller enabled the formation of a multiagent system characterized by a leader-follower structure. Such a control strategy allowed each agent to have the same output even with a different number of inputs, facilitating the system task planning.

Extensive research efforts were also directed toward modeling unwanted human control behavior in transport systems, particularly rotorcraft–pilot coupling (RPC) (Pavel *et al.*, 2015). For evaluating human–rotorcraft interaction in aspects such as comfort and handling qualities, some modeling efforts found in the literature were directed toward studying the dynamical behavior of the human body. Understanding such body dynamics, in this case, the upper body is fundamental to identifying potentially dangerous nonlinearities in RPC. These approaches vary significantly and can be classified several main categories, such as finite element models (FEMs), multibody dynamics (MBD), and lumped parameter models (LPM).

LPMs are composed of elementary mechanical subsystems, such as lumped masses and viscoelastic elements with linear or nonlinear properties. In the linear case, parameters are relatively easy to identify, with a low associated computational cost, and can be easily tuned to fit the biomechanical characteristics of a specific subject. However, in LPMs where nonlinear viscoelastic elements are used, the cost of identifying its characteristics may increase, depending on the applied force or displacement. Munyazikwiye *et al.* (2018) used a piecewise LPM as an analytical tool to perform a preliminary analysis of vehicle crashworthiness in order to reduce the time required to assemble and tune FEMs and perform a nonlinear finite element analysis in crash testing. In the proposed LPM, the spring and damping coefficients are defined as piecewise linear functions of input displacement and velocity. Lumped parameters nonlinear models are also present in works such as that by Payne and Stech (1969), in which a one-degree-of-freedom model was applied for analyzing human body dynamic response during a helicopter landing. In works such as those by Abbas *et al.* (2010) and Xu *et al.* (2008), previous state-of-the-art linear models were optimized using a genetic algorithm to capture the nonlinear effects of passengers' dynamic response when subjected to vibrations.

Zanoni *et al.* (2020) designed a multibody model of the upper body by connecting a model of the pilot's arms to a model of the spine. Such a spine model, as well as the scaling procedures, was used for studying seat-to-head transmissibility. This coupled spine-arms model can be used to evaluate the biodynamic response of the human operator in terms of involuntary motion induced on the control inceptors, including the related nonlinearities.

FEMs have been successfully used in recent research to represent human body behavior during an impact, often in relation to injury risk prediction and vehicle safety. The total human model for safety (THUMS) is a famous finite element human body model intended for injury analysis (Iwamoto *et al.*, 2002); it has been used in association with a model of a vehicle's internal structure, with the purpose of simulating human body kinematics in response to a large impact in a car crash. The geometries of the structurally complex human body parts, including the head, torso, ligaments, joints, and internal organs, are represented by finite element meshes, and their impact responses have been studied separately. Moreover, in relation to transport safety, within the context of the European project "Human model for safety two" (HUMOS2) (Vezin and Verriest, 2005), human body numerical body models were constructed in order to create a database able to represent the European population with high fidelity. Portions of HUMOS2 models have been used in many research efforts, such as by von Merten (2006) for thoracic accidents and by (Toma *et al.*, 2010) for head injuries in motorcycle crashes. Another example of FEM used to provide kinematic and kinetic data of the human body in a computationally efficient way has been proposed by Schwartz *et al.* (2015) at the Global Human Body Models Consortium (GHBMC).

4.6 Final Considerations

The presented modeling research efforts of nonlinear dynamics in human–machine interaction successfully captured many aspects of the human learning process, information processing, and control action. From the classical control-theory approach of dual-loop control to the more recent machine-learning techniques, many advances have been made in identifying the sources of nonlinearity in human control behavior and in implementing models able to transfer such ability to the controlled machines. Modeling and data-driven techniques were presented in a human-centered way in order to show how they succeeded in representing different aspects of the human as a controller. For instance, the decision-making process directed toward achieving an internal goal is well described by RL approaches, while optimal control models of the neuromuscular system or biodynamical models are most useful for nonlinear dynamics deriving from human body actuation districts or from their coupling with the controlled element. Moreover, data-driven techniques associated with control systems were analyzed in relation to nonlinearities that derive from the controlled element dynamics and/or the external environment. As proved by the discussed man-machine systems, algorithms can be combined to increase the level of autonomy and the usability of machines even in complex scenarios such as connected vehicles, automatic lane changes, teleoperation, or remote control of rotorcraft.

This is done while acting in an environment surrounded by humans, with consequent potential issues regarding safety and adding unexpected physical interaction that requires a level of adaptability, which is typical of human beings and constitutes one of the reasons that motivated such modeling efforts. Despite the successes concerning classical control theory models discussed in the first sections, modern machine learning frameworks struggle to capture the physiological context relying upon the human learning process. Neural networks and algorithms based on RL or optimal control paradigms still have almost a black-box approach to what concerns this aspect. Advances in understanding the human brain are still a challenge that motivates many research activities.

However, this way of combining different machine learning and model-based techniques seems to be perfect for our human–robot interaction application scenario. In this chapter, we discussed nonlinear models of human–machine dynamics in general without thinking of a specific human control aspect to model and predict; in the following chapter, we will focus our analysis on how these kinds of approaches can be used specifically in human intention estimation. Once again, we will propose different applicative use cases in which the discussed modeling frameworks are used.

Chapter 5

Human Motion Prediction

As we saw in the last chapter, the evolution of autonomous systems has led to continuous integration into human environments. These systems are becoming increasingly essential to perceive, interpret, and predict human movements.

We explored the models that in the literature have been used to capture human nonlinear dynamics, which are crucial to characterize their behavior when interacting with a machine. We now move our focus from the general discussion on nonlinear control dynamics to the specific application of our interest: human motion prediction.

Having predictive capabilities concerning human motion is crucial in various applications, from self-driving vehicles and service robots to advanced surveillance. Here, we present a comprehensive survey on this topic, proposing, as usual, different application domains in which such aspects have been successfully developed.

5.1   Introduction

Human trajectory prediction encompasses diverse applications, including pedestrian path estimation for autonomous vehicles, force-assisted movement in industrial robots, and real-time motion adaptation in assistive devices. Apart from the aforementioned shift of our focus on motion prediction, this chapter also aims to emphasize how these approaches can be transformed into actionable technological solutions.

For autonomous systems to coexist effectively with humans, they need a robust understanding of human motion. This ability, which includes recognizing and analyzing motion patterns, allows systems to anticipate how situations will evolve, supporting enhanced planning and interaction capabilities, like active perception and model predictive control. As a result, human motion prediction has gained significant traction in domains such as autonomous driving, robotics, and surveillance.

The challenge of human trajectory prediction arises from the complexity of human behavior, influenced by individual goals, social norms, other agents, and environmental factors like physical layout and affordances. Many of these elements aren't directly observable, requiring interpretation from sensory data or inferred context. Predictions must be accurate and quick enough for real-time use for practical applications.

Human motion can take many forms, including walking, using assistive devices, or driving. This survey focuses specifically on predicting pedestrian paths in 2D ground-level scenarios and covers relevant research on cyclists and vehicles. We structure the field based on different modeling methods and contextual considerations, discussing each category's strengths and limitations and identifying areas where research could advance further.

We refer to dynamic agents such as people, vehicles, or robots as "agents," while "target agents" are the objects for which predictions are made. The term "trajectory" refers to a path with a timing or velocity element, and prediction time frames vary, with "short-term" covering a few seconds and "long-term" up to 20s ahead.

Trajectory prediction is a concept widely used in various technological fields, such as robotics, artificial intelligence, and autonomous systems, and plays a crucial role in human–robot interaction. In robotics, trajectory prediction algorithms enable machines to navigate dynamic environments by forecasting the future paths of objects or humans. Similarly, in automotive technology, predictive models help autonomous vehicles anticipate the movements of other vehicles or pedestrians, leading to safer decision-making processes. These concepts are being progressively adapted to rehabilitation settings, where robots assist humans in completing predefined motor tasks.

Moving from the general to the particular, also in the rehabilitation field, predicting the upper limb's trajectory has been used to anticipate human motion, both with model-based and data-driven techniques, often combined. With respect to the other human–robot interaction application scenarios, rehabilitation presents unique challenges due to the patient's variability in motor control recovery. Trajectory patterns can be affected by neuromuscular deficits, compensatory strategies, and the degree of spasticity, making accurate prediction vital for robotic systems designed to provide rehabilitative care. By leveraging predictive models, robotic systems can adjust assistance levels based on the anticipated movement, thus ensuring personalized rehabilitation protocols that evolve with the patient's progress.

Ultimately, we explore the applicability of trajectory prediction in upper-limb rehabilitation with specific reference to a novel prototypical robotic platform that can be used for lower-limb and upper-limb rehabilitation of post-stroke patients. The system is an end-effector Cartesian robot, which guides the patient and anticipates its motion during compensatory tasks, such as reaching or grasping movements.

Previous surveys have examined specific aspects of human motion prediction in fields like robotics, intelligent vehicles, and computer vision. Studies have explored socially aware navigation for robots, categorizing approaches by comfort, naturalness, and social norms. Other surveys, particularly in autonomous driving, have categorized methods based on factors like motion intent, dynamics-based modeling, and social or environmental awareness. Our review seeks to build upon and unify these approaches, proposing a broader taxonomy that recognizes the influence of contextual information and provides a structured foundation for comparing different methods.

We categorize motion prediction approaches based on modeling techniques and contextual information:

- **Modeling approaches**: Techniques vary in how they represent motion, whether using physical laws, learned patterns, or reasoning about an agent's goals. Physics-based methods follow explicit dynamics, pattern-based methods derive insights from historical data, and planning-based approaches involve predicting paths based on anticipated goals.
- **Contextual cues**: Additional information about an environment or social context can influence predictions, such as spatial layout, semantics, or the behavior of surrounding agents. Contextual awareness can refine predictions and improve system responsiveness.

By structuring our review across both modeling methods and context, we highlight where certain approaches excel and where further research could expand capabilities, especially in capturing the nuances of human behavior in complex environments.

5.2 Classification Rules

Some works in the survey span multiple categories within the taxonomy, especially those integrating several approaches. In such cases, classification is based on the following specific guidelines:

(1) Methods are generally classified by their primary modeling technique rather than their inference method.
(2) Approaches incorporating elements from different categories are classified according to their main modeling approach.
(3) Direct learning of motion from data is classified as pattern-based, whereas imitation learning with policy reasoning falls under planning-based.

Additionally, each work is grouped by contextual cues: those involving the target agent, static cues, and dynamic cues.

5.3 Physics-Based Approaches

Physics-based models predict human motion using a predefined dynamic model grounded in Newtonian mechanics. Typically, the dynamical model takes the form

$$\dot{\mathbf{s}}_t = f(\mathbf{s}_t, \mathbf{u}_t, t) + \mathbf{w}_t,$$

where $\mathbf{u}_t$ represents the control input and $\mathbf{w}_t$ represents process noise. Motion prediction in this context requires estimating both $\mathbf{s}_t$ and $\mathbf{u}_t$ based on observed cues.

These models are frequently used in tracking and control, covering both maneuvering and non-maneuvering targets. They fall into two categories: (1) single-model approaches, which use one specific model for the dynamics, and (2) multiple-model (MM) approaches, which involve several dynamic modes.

5.3.1 *Single-model approaches*

5.3.1.1 *Early works and basic models*

In early approaches, target states are represented by position, velocity, and acceleration parameters. Common models include constant velocity (CV), constant acceleration (CA), and coordinated turn (CT) models, which are effective for limited prediction horizons and low motion uncertainty.

Dynamic models incorporating forces offer enhanced accuracy for longer-term predictions. Autoregressive models (ARMs), which utilize historical state data, have also been applied to motion tracking.

For constrained environments like roads, models often include map-based cues. In other settings, grid-based or graph-based methods are used.

5.3.1.2 *Models with dynamic environment cues*

In social contexts, local interactions are often modeled using frameworks like the social force model, adapted to improve short-term pedestrian tracking and goal estimation. For vehicle applications, models like the intelligent driver model (IDM) predict longitudinal motion and have been used for intent inference in urban traffic scenarios.

5.3.2 *Multi-model approaches*

Agent motion is often too complex to be captured by a single dynamic model. While combining map details and interactions with multiple agents can improve flexibility, limitations persist. A common strategy involves using a set of prototypical motion modes, such as linear motion, turning, or accelerating. These modes, when sequenced, describe intricate motion patterns. Since other agents' motion modes aren't directly observable, methods are needed to handle this uncertainty, typically through multi-model (MM) and hybrid estimation approaches.

MM methods represent the state as a hybrid system, including continuous dynamics and discrete motion modes. Key components include a model set, a strategy for managing discrete uncertainties (often under a Markov assumption), a recursive estimation for continuous values, and a fusion mechanism. MM methods enhance motion prediction by accounting for agent interactions and map-based context.

A basic MM approach involves a mixture of linear dynamics (e.g., turning or moving straight) constrained by the map. Interactive multiple model (IMM) filters are widely used in MM-based tracking and prediction, such as in vehicle and pedestrian state estimation. For example, vehicle state estimation can alternate between constant acceleration and simplified bicycle dynamics. Pedestrian prediction often combines simple motion models with additional intentions inferred from situational factors like head orientation.

More advanced MM techniques include dynamic Bayesian networks (DBNs), allowing flexible conditioning on agent context. For instance, DBN models have been applied to long-term multi-vehicle prediction in mining vehicles, integrating contextual features like agent distance and location cues. Similarly, in urban traffic, DBNs can model pedestrian paths by incorporating head orientation and curb distance as contextual factors.

Hybrid approaches, based on stochastic reachability theory, predict agent motions by computing reachable sets. These approaches consider agents as systems with multiple motion modes and infer their behavior through probabilistic or deterministic future states, such as in a set of allowable actions for human prediction.

5.4 Motion Prediction in Human–Robot Interaction

Understanding human motion is essential for intelligent systems to interact effectively and coexist with humans. It involves key components like representation, perception, and motion analysis. Prediction plays a crucial role in this process, as anticipating how a scene with multiple agents will evolve over time allows systems to act proactively. This is beneficial for tasks such as active perception, predictive planning, model predictive control, and human–robot interaction. Consequently, human motion prediction has gained considerable attention across various fields, including self-driving vehicles, service robotics, and advanced surveillance systems.

Accurately predicting human motion is challenging due to the complexity of human behavior, which is influenced by numerous internal and external stimuli. Motion behavior may be driven by intentions, the presence and actions of nearby agents, social dynamics, cultural norms, or environmental factors like physical layout and semantics. Many of these influencing factors are not directly observable, requiring inference from noisy sensory inputs or contextual modeling. Additionally, effective motion prediction systems must be both robust and capable of operating in real time.

At a high level, the motion prediction problem can be broken down into the following three main components (as shown in Figure 5.1):

- **Stimuli**: These are the internal and external factors influencing an agent's motion, such as goals, intentions, or interactions with other agents and the environment. Typically, prediction systems rely on partially observed trajectories or sequences of state observations like position, velocity, joint angles, or other attributes. These observations are often tracked through systems that assume correct identification over

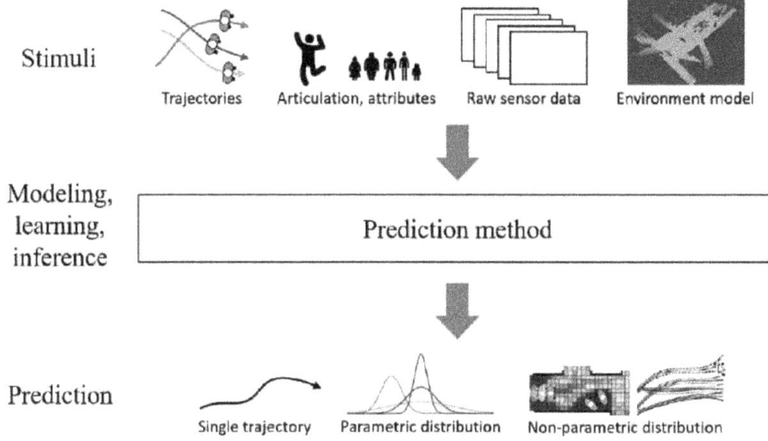

Fig. 5.1. Typical elements of a motion prediction system: internal and external stimuli that influence motion behavior, the method itself, and the different parametric, non-parametric, or structured forms of predictions.

time. Inputs may also include contextual data such as scene geometry and environmental semantics or cues from surrounding entities. Some methods rely on raw sensor data in an end-to-end fashion.

- **Modeling approach**: Prediction methods vary in how they represent, parameterize, and learn to model human motion. This text focuses on identifying functional categories, underlying similarities, shared assumptions, and best evaluation practices across the growing literature in this domain.

- **Prediction**: Methods differ in how they produce predictions—ranging from parametric or non-parametric approaches to structured representations. These include Gaussian distributions over agent states, probability maps, single or multiple trajectory samples, or motion patterns captured through graphical models.

In this context, the term "agent" refers to dynamic entities of interest, such as robots, pedestrians, cyclists, or vehicles. The "target agent" is the specific entity for which motion is being predicted. We assume that the target agent's behavior is not erratic but rather goal-oriented, seeking an optimal or near-optimal outcome. Without this assumption, the problem of motion prediction would become exceedingly complex or ill-defined. A "path" is defined as a sequence of (x, y) positions, while a "trajectory" combines this path with timing information or a velocity profile. Predictions can be short term (1–2 s) or long term (up to 20s).

Formally, the state of an agent at time t is denoted as $\mathbf{s}_t$, while the action taken by the agent at time t is represented by $\mathbf{u}_t$. The observation of the agent's state at time t, $\mathbf{o}_t \in \mathcal{O}$, represents the perceptual inputs, and ζ denotes trajectories. A history of states, actions, or observations over a time range t to T is denoted by the subscript t.

Human motion takes various forms, ranging from full-body articulation and gestures to facial expressions or spatial movement through walking, using a mobility device, or driving. This discussion focuses specifically on human motion trajectory prediction, particularly in 2D for ground-level pedestrians, but also considers cyclists and vehicles. The prediction of video frames, articulated motion, or human actions is outside the scope of this work, though many of the same motion modeling principles apply.

Many research efforts have been carried out in such directions in many application domains, such as the following:

- Service robots: Mobile service robots increasingly operate in open-ended domestic, industrial, and urban environments shared with humans. Anticipating the motion of surrounding agents is an essential prerequisite for safe and efficient motion planning and human–robot interaction. Limited onboard resources for computation and first-person sensing make this a challenging task.
- Self-driving vehicles: The ability to anticipate the motion of other road users is essential for automated driving. Similar challenges apply in the service robot domain. However, they are more pronounced given the higher masses and velocities of vehicles, resulting in more considerable harm that can be inflicted, especially toward vulnerable road users (i.e., pedestrians and cyclists). Furthermore, vehicles must operate in rapidly changing, semantically rich outdoor traffic settings and need hard real-time operating constraints. Knowledge of the traffic infrastructure (location of lanes, curbside, traffic signs, traffic lights, and other road markings, such as zebra crossings), and traffic rules can help predict motion.
- Surveillance: Visual surveillance of vehicular traffic or human crowds relies on accurately tracking many targets across distributed networks of stationary cameras. Long-term motion prediction can support a variety of surveillance tasks such as person retrieval, perimeter protection, traffic monitoring, crowd management, or retail analytics by further

reducing the number of false-positive tracks and track identifier switches, particularly in dense crowds or across non-overlapping fields of views.

Lasota *et al.* (2017) reviewed the state-of-the-art safe human–robot interaction, analyzing four aspects: safety through control, motion planning, prediction, and psychological factors. The work on such topics included wheeled robots, manipulator arms, drones, and self-driving vehicles.

The literature on human motion prediction is divided into methods based on goal intent or motion characteristics.

Goal intent techniques infer an agent's goal and predict a trajectory the agent will likely take to reach that goal. For goal estimation, the observed people trajectories can be, for example, represented in a sparse topological map of the environment (Elfring *et al.*, 2014). Each node of the map encodes a state–destination pair, and the goal inference using the observed trajectory is carried out in a maximum-likelihood manner. The approach proposed by Oli *et al.* (2013), instead, defined a robot motion planner operating in social spaces as an interacting agent affected by social forces. Each human was flagged as either aware or unaware of the robot, which defined the repulsive force the robot exerts on that person. Such awareness was inferred using visual cues (gaze direction and past trajectory). In order to achieve more realistic behaviors, several extensions to the social force model have been proposed (Yan *et al.*, 2014), where it was presented as a model that embeds social relationships in the linear combination of predefined primary social effects (attraction, repulsion, and non-interaction).

The motion predictor maintains several hypotheses over the social modes in which the users are involved. Karamouzas and Overmars (2010) proposed a velocity-based approach to model and simulate human collision avoidance predictive behavior. In particular, each agent (human) was modeled to adapt their route as early as possible, trying to minimize the number of interactions with others and the energy required to solve these interactions. To this end, an evasion force that depends on the predicted point of collision and the distance to it is applied to each agent.

The motion characteristics group of approaches, on the other hand, does not rely explicitly on goals and makes use of observations about how humans move and plan natural paths; a domain of study has been the already mentioned video surveillance, in which agents, their trajectory, and behavior need to be observed and predicted. Morris and Trivedi (2011) analyzed the methods for trajectory learning and analysis for visual surveillance. In particular, a three-stage hierarchical modeling process characterized activities at multiple levels of resolution and has been used to develop, classify, and predict them in the future and detect abnormal behavior. Hirakawa *et al.* (2018) surveyed video-based methods for semantic feature extraction and human trajectory prediction. The literature is divided based on the motion modeling approach into Bayesian models, energy minimization methods, deep learning methods, inverse reinforcement learning (IRL) methods, and other approaches.

5.5   Trajectory Prediction in Robotic Rehabilitation

Trajectory prediction and human intention estimation methods have also found extensive applications in rehabilitation. Wei *et al.* (2024) reviewed model-free (MF) and model-based (MB) methods for motion prediction in rehabilitation systems characterized by continuous human–robot interaction.

According to the prediction methods used in the analyzed state-of-the-art studies, continuous regression can be divided into MB and MF approaches (Figure 5.2). Hence, given the current research landscape and the fact that rehabilitation robots are continuously controlled, we will focus on studies using MB and MF methods for continuous upper limb motion intention (i.e., joint kinematics and dynamics) prediction.

MF approaches mainly encompass machine learning (ML) and deep learning (DL) methods. Traditional ML methods may require manual feature extraction and selection from pre-processed surface electromyography (sEMG) signals. At the same time, DL can automatically extract advanced features from sEMG and utilize the neural network's potent fitting capacity to approximate the highly nonlinear relationship between features and motion intentions, thereby avoiding the reliance on optimal feature

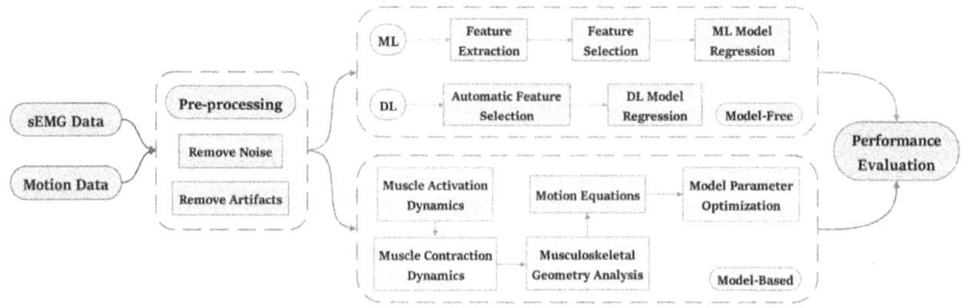

Fig. 5.2.   Motion intention modeling process as studied by Wei *et al.* (2024).

sets based on empiricism similar to ML (Xiong *et al.*, 2021). Although end-to-end MF methods are convenient to train and quick to deploy, their inherently "black box" nature may overlook the physiological causal relationships between input and output data, consequently struggling to generalize beyond the training data and risking overfitting (Pizzolato *et al.*, 2019; Scibilia *et al.*, 2024b; Sitole and Sup, 2023).

In contrast, MB methods with inherent physiological causality can convert sEMG signals to muscle-tendon forces according to the neural-physiological mechanisms of muscle activation and contraction dynamics before predicting joint kinematics and dynamics from joint torques using musculoskeletal (MSK) geometry and Newtonian motion equations. Additionally, most MB studies employed Hill-type MSK models with series elastic (SE), contractile element (CE), parallel elastic (PE), and viscoelastic (VE) components. However, due to physiological differences among patients, the generic proportional Hill model derived from extensive cadaveric specimens can lead to significant prediction errors. Therefore, parameter optimization for subject-specific MSK models in the final stage of MB methods is necessary to achieve precise predictions (Pizzolato *et al.*, 2017).

Focusing on upper limb movements, we begin by discussing techniques that use one or multiple joint models and, as we will see, often integrate them with data-driven algorithms to predict the subject's motion. Research on upper limb joints has predominantly focused on the elbow. At the same time, the shoulder has received less attention due to its anatomical complexity, the diversity of multi-degree of freedom (DoF) movements, and the inherent difficulties in acquiring sEMG signals. One study (Aung *et al.*, 2015) addressed these challenges by integrating a muscle activation model with electromechanical delay (EMD) and extreme learning machine (ELM), enabling real-time prediction at a low latency of 32 ms, even during random movement speeds.

For the elbow joint, different modeling approaches have been explored. One study (Pang *et al.*, 2015) used a simplified Hill model that included only the CE element and combined it with a state-switching model for real-time prediction. Despite achieving real-time performance, this method showed time delays and lacked robustness when minor joint angle variations occurred. By contrast, other research employed the complete Hill model, which includes the CE, PE, and SE elements (Han *et al.*, 2015a), or opted for a rigid-tendon version incorporating only the CE and PE elements (Desplenter and Trejos, 2018). The latter approach, however, failed to account for muscle stiffness variations, leading to significant torque prediction errors. To address these shortcomings, some researchers (Li *et al.*, 2018) enhanced the Hill model using a genetic algorithm (GA) combined with a short-range stiffness (SRS) model, enabling concurrent predictions of joint angles and time-varying stiffness. Another solution was to optimize the Hill model with additional physiological parameters, improving robustness across varied movement loads (Li *et al.*, 2019a), while others used nonlinear least squares for model optimization.

Regarding wrist joints, several studies employed the rigid-tendon Hill model to predict joint behavior. Specifically, sensitivity analysis and GA optimization were used in some studies to balance the model's predictive accuracy and computational complexity (Zhao *et al.*, 2023b, 2020). These efforts addressed the issue of oversimplification, which often leads to overestimated parameters and neglects individual subject specificity. One experiment based on mirrored bilateral motion demonstrated no significant statistical difference in performance between ipsilateral and contralateral sides, confirming the method's robustness (Zhao *et al.*, 2023b). Because the supinator muscle is difficult to measure via sEMG due to its deep-seated location, other studies employed non-negative matrix factorization (NMF) for virtual muscle co-activation to replace pronator and supinator activations in the Hill model (Li *et al.*, 2023; Wang *et al.*, 2021b). This approach outperformed linear regression (LR) and artificial neural networks (ANNs) based on time-domain (TD) and muscle synergy (MS) features, showing greater robustness across various upper limb postures.

As evident, the general effort is to improve models' performances by enriching them with different models, algorithms, and optimization techniques. Many studies went in this direction by combining MB and MF approaches (Pang and Guo, 2013; Yang *et al.*, 2020). One way that has been explored is to replace the traditional MSK geometric analysis and motion equations with polynomials to approximate the relationship between muscle-tendon force and elbow angles (Pang and Guo, 2013). Another developed a state space (SS) model optimized through an extended Kalman filter (EKF) by integrating the Hill model with TD features, allowing real-time closed-loop estimation (Han *et al.*, 2015b). A comparison between the Hill model and a backpropagation neural network (BPNN) for predicting elbow isometric contraction force concluded that, while the BPNN outperformed the Hill model due to the linear relationship between joint forces and sEMG, it lacked interpretability (Yang *et al.*, 2020). This highlighted the complementary benefits of combining MB and MF methods. For instance, radial basis function neural networks (RBFNNs) have optimized muscle activation factors, reducing conversion bias between sEMG and muscle activation and improving adaptation to individual characteristics (Xu *et al.*, 2021).

Concerning wrist modeling, one research study utilized the BPNN to identify different motion phases based on muscle activation (Zhang *et al.*, 2016). Then, it employed an MSK model optimized via Bayesian LR for low-latency, real-time joint force prediction (Zhang *et al.*, 2022a). The findings indicated that Bayesian LR was more efficient and straightforward than GA-based optimizations. Another approach introduced physical MSK knowledge as a soft constraint in the loss function of a convolutional neural network (CNN), outperforming methods such as support vector regression (SVR), ELM, multi-layer ELM (ML-ELM), and traditional CNN architectures (Zhang *et al.*, 2023b). This model required less training data, converged faster, and had a simpler architecture. Building on this work, transfer learning techniques were applied by reusing pre-trained CNN parameters and only updating the fully connected layer. This method demonstrated excellent convergence rates, generalization, and minimal need for individual data, enabling rapid transfer learning (Zhang *et al.*, 2023a).

5.6 From Modeling to Controlling Frameworks: Model Predictive Control

In human–robot interaction scenarios such as those analyzed in the previous sections, where continuous physical collaboration between humans and robots is key, the possibility of translating the models' predictions into a proper control framework has become an essential tool for enhancing robot responses to human actions. As evidenced by the analysis we have carried out until this point, hybrid approaches combining classical models with machine learning frameworks have gained prominence due to their ability to generalize over unstructured data while leveraging partial system knowledge (Gupta and Smith, 2022). This concept is also valid from the control frameworks developed from such models.

Predictive control techniques have found widespread application in HRI scenarios, where anticipating human behavior in real time is critical for safe and effective collaboration. In such human-in-the-loop systems, the robot must accurately predict future human actions based on sensory feedback and historical data, allowing it to adjust its behavior in response to dynamic and often unpredictable human inputs. This approach is particularly valuable in tasks where variability is inherent, such as those requiring adjustments based on force or motion. Among these techniques, *model predictive control* (MPC) stands out as a widely studied and effective method in robotics (Wei and Lu, 2018). By leveraging a model of the system, MPC predicts future states and computes control inputs that optimize a given performance criterion. This framework has been particularly beneficial in enabling robots to react to human intentions by forecasting movements or force applications.

One significant application of MPC lies in industrial robotics, where its integration enables precise interaction control. For example, MPC frameworks are employed to optimize both the safety and efficiency of tasks like suturing, which involves intricate, dynamic motion planning (Caianiello *et al.*, 2024). These frameworks demonstrate how predictive modeling can ensure smooth execution by minimizing the risk of collision or task failure. Another example that can be carried out in the industrial context is related to tasks like assembly or material handling. In these environments, humans provide dynamic and variable inputs, such as applied forces or trajectory changes, which the robot must anticipate and compensate for (Wei and Zhang, 2018). Traditional control techniques that assume stationary or predictable environments are ill-suited for such tasks. Instead, MPC-based systems provide real-time predictions, allowing robots to adapt continuously to human actions, even in noisy or uncertain conditions.

Moreover, the integration of MPC in collaborative scenarios goes beyond physical control. By employing contact models, robots are better equipped to manage dynamic interactions, ensuring a natural response to human-guided movements. For example, contact-based MPC approaches have been shown to improve the fluidity and responsiveness of robots during interaction-heavy tasks (Zube *et al.*, 2016).

In rehabilitation, MPC frameworks support advancements in personalized systems. These systems infer user intent, such as balance control, and adapt their assistance dynamically, thus ensuring safety and comfort during recovery. Using inverse MPC techniques, researchers have successfully developed methods to detect and respond to subtle changes in user behavior (Ramadan *et al.*, 2019).

Additionally, nonlinear MPC models are being utilized to predict human center-of-mass trajectories, which are vital for ensuring stability in tasks like gait training and balance support. These models integrate biomechanical data to create adaptive robotic systems that align seamlessly with individual needs (Joos *et al.*, 2018).

Another critical innovation is the use of adaptive impedance control integrated with MPC. By adjusting the robot's stiffness dynamically based on environmental and interaction feedback, systems ensure smoother transitions and safer operations (Xiao *et al.*, 2024). This technique is particularly useful in scenarios requiring delicate manipulation or force compliance.

In human-following robotics, dual closed-loop MPC strategies respect social space constraints, enhancing both safety and user comfort. These systems predict and adapt to the dynamic behaviors of humans in shared environments, creating robots that operate harmoniously alongside people (Peng *et al.*, 2024).

5.7 MPC Using Hybrid Data-Driven and Model-Based Techniques

However, purely model-based approaches face limitations, particularly in handling nonlinearity and high-dimensional data. This has driven the integration of machine learning techniques into predictive control frameworks. For instance, ANNs have been combined with MPC to enhance its predictive capabilities in scenarios where the dynamics of human–robot interaction are too complex for traditional models (Raffin and Stulp, 2019). ANNs, with their ability to approximate nonlinear functions, can be trained on historical data to model the intricate relationships between human inputs and robot responses. This is particularly valuable in applications requiring robots to predict human intentions from limited and noisy sensory data.

One notable example is the use of *recurrent neural networks* (RNNs) and *long short-term memory* (LSTM) networks in predictive control tasks with time-dependent data (Hochreiter and Schmidhuber, 1997). These networks capture the temporal correlations in human actions, enabling robots to better anticipate future movements. In assistive robotics for rehabilitation, RNNs have been used to model the interaction forces between patients and robotic arms, predicting patient movements based on past motion data (Gao and Sun, 2020). Such predictive models improve robot assistance while ensuring safer and more responsive interactions.

Hybrid control systems, which combine data-driven techniques with classical model-based methods, represent a promising direction for improving predictive accuracy in human–robot interaction. These systems leverage the adaptability of machine learning and the robustness of structured models (Nguyen and Luo, 2021). For example, MPC can serve as a structured control framework, while learned models like ANNs or reinforcement learning adapt to human interaction nuances.

In physical human–robot collaboration (pHRC), reinforcement learning has been employed to train robots to adapt to human force patterns and motions over time, optimizing safety and performance (Kormushev *et al.*, 2010). While these methods often require extensive training data and computational resources, the incorporation of predictive models such as MPC provides a foundation for guiding real-time adjustments.

Among data-driven methods, the *nonlinear autoregressive moving average with exogenous inputs* (NARMAX) model has gained attention for its ability to capture complex, nonlinear dynamics using past data. In human–robot interaction, NARMAX models are effective for forecasting human-applied forces in real-time, making them suitable for tasks requiring accurate, short-term predictions (Billings, 2013). By integrating autoregressive and moving average components, NARMAX captures interaction dynamics while maintaining computational efficiency for real-time applications.

MPC's role in integrating predictive models into robotic systems extends to optimizing trajectories and real-time adjustments. In human–robot collaboration, robots equipped with MPC can adapt impedance in response to anticipated human motion, ensuring safety and efficiency in shared workspaces. Recent advancements explore hybrid frameworks integrating NARMAX and other machine learning techniques to enhance predictive accuracy while maintaining computational tractability (Scibilia *et al.*, 2024a).

Future research in MPC-based human–robot interaction will continue to explore hybrid approaches combining MPC with learning-based methods. For instance, teaching-by-learning frameworks allow robots to personalize control strategies by understanding user-specific movements and preferences over time (Safavi and Zadeh, 2017). This combination of prediction and adaptation forms the basis for truly collaborative and intuitive robotic systems.

Finally, applying MPC in redundant manipulators highlights its role in human–robot collaboration. These systems optimize motion planning by dynamically adjusting trajectories based on the shared task context, improving efficiency and reducing the cognitive load on human operators (Cai *et al.*, 2024).

5.8 Final Considerations

The discussions presented in this chapter highlighted the growing relevance of modeling techniques applied to human motion prediction in many kinds of autonomous systems, including those characterized by closer human–robot physical interaction. In particular, from all the discussed models, the most encouraging results and ease of use have been observed when there was an integration between data-driven and model-based techniques.

With the increasing demand for safe and effective human–robot collaboration, research efforts were not limited to modeling human intentions and movements but translated them into proper control methods able to use such information for enhancing the system's adaptability. The perfect example of this concept is MPC, which can be considered the most common and effective way to use hybrid modeling frameworks to enable robots to precisely anticipate and respond to human actions.

In the context of continuous physical interaction, where only partial system knowledge is available, the combination of machine learning frameworks with ARMs, such as NARMAX, offers a promising pathway. These approaches effectively handle noisy and unstructured data while maintaining computational efficiency, making them ideal for real-time applications.

Building on these state-of-the-art techniques, in the following chapter, we will propose a novel NARMAX-based modeling framework for estimating human force in human–robot collaboration, focusing both on model development and on how such a system can be used in practical scenarios to enhance robot controllers' performances.

Chapter 6

NARMAX Model for Human Force Estimation

6.1 NARMAX Models

Nonlinear autoregressive models with moving average and exogenous input (NARMAX) was proposed for the first time by Chen and Billings (1989) Many state-of-the-art research efforts have been widely used to identify nonlinear systems with partially unknown features with good results. As may be deducible from the acronym, NARMAX models derive from the mostly known linear ARMAX models and their variations like AR, ARX, ARMA, and ARIMA.

Billings (2013) considers NARMAX not only as the name of a model but also as a proper philosophy of nonlinear system identification that consists of the following five steps:

- **Structure selection**: Identify the terms present in the model.
- **Parameter estimation**: Estimate its coefficients.
- **Model validation**: Check eventual errors and biases in the model.
- **Prediction**: Forecast model output in several steps in the future.
- **Analysis**: Analyze the dynamical properties of the system.

The first step, structure detection, constitutes the crucial part of nonlinear model identification. While in the linear case determining model order is relatively easy and is often between the first and third orders (making the computations even faster and more efficient), things are far more complicated and computationally demanding in the nonlinear case, where the number of candidate terms can rise easily if we want, for example, to approximate lag terms with polynomial expansions (Nepomuceno and Martins, 2016).

Applications of NARMAX in robotics have found increasing fortune in the last decade due to their flexibility and capability of characterizing system response even in the presence of unknown or highly nonlinear dynamics. For instance, Kyriacou *et al.* (2006) demonstrated the use of these models in visual task identification, where NARMAX captured dynamic responses in rapidly changing environments. Similarly, Hussain *et al.* (2019) utilized NARMAX to optimize control strategies in human–robot collaboration, showcasing its robustness in adapting to human inputs in real time. These studies emphasize the versatility of NARMAX, not only as a modeling tool but also as a framework for developing predictive and adaptive control systems.

Let us consider the experimental setup proposed in Chapter 3, in which our model has to predict the force actuated by a human subject into the robot end-effector when an external reference has been applied to the system. If we indicate the force signal we want to predict as y and the external reference (or exogenous input) as x, the model would be

$$y(t) = f(y(t-1), \ldots, y(t-n_a), x(t-d), \ldots, x(t-d-n_b),$$
$$e(t-1), \ldots, e(t-n_c)), \tag{6.1}$$

where n_a, n_b, and n_c indicate the maximum lags for system output, system input, and noise, respectively, and define the order of the model, while d indicates an additional transport delay between input and output.

6.1.1 *Polynomial approximation*

The nonlinear function $f(\cdot)$ can be approximated using different strategies; we will focus on polynomial algorithms and neural networks. The first one aims to find a piecewise linear equivalent of the nonlinear function. The following equation can describe a polynomial NARMAX model with asymptotically stable equilibrium points (Lacerda *et al.*, 2020):

$$y_t = \sum_0 + \sum_{i=1}^{p} \Theta_y^i y_{k-i} + \sum_{j=1}^{q} \Theta_e^j e_{k-j} + \sum_{m=1}^{r} \Theta_x^m x_{k-m} + \sum_{i=1}^{p}\sum_{j=1}^{q} \Theta_{ye}^{ij} y_{k-i} e_{k-j}$$

$$+ \sum_{i=1}^{p}\sum_{m=1}^{r} \Theta_{yx}^{im} y_{k-1} x_{k-m} + \sum_{j=1}^{q}\sum_{m=1}^{r} \Theta_{ex}^{jm} e_{k-j} x_{k-m}$$

$$+ \cdots + \sum_{m_1=1}^{r} \cdots \sum_{m_l=m_{l-1}}^{r} \Theta_{x^l}^{m_1,\ldots,m_2} x_{k-m_1} x_{l-m_l} \tag{6.2}$$

where $\sum_0$ and all the terms with Θ indicate constant parameters and l is the degree of polynomial nonlinearity. Polynomial algorithms will select a subset of the terms of equation (6.2), also called regressors, which will minimize the error between estimated $\hat{y}(t)$ and real $y(t)$. Examples of these kinds of algorithms are the forward regression least squares (FROLS) (Aguirre *et al.*, 1998), meta-model structure selection (MetaMSS) (Junior *et al.*, 2021), accelerated orthogonal least squares (AOLS) (Hashemi and Vikalo, 2018), and entropic regression (AlMomani *et al.*, 2020; Kraskov *et al.*, 2004). Most of them use information theory for order selection and select the best regressors and least-squares-based algorithms, in the case of FROLS and AOLS, for parameter estimation. This kind of strategy, as explained by Lacerda *et al.* (2020), has a critical aspect in required computational time that can become very high due to the fact that the possible model structures to be tested are 2^{n_r}. Here, n_r is the number of candidate regressors that depends on the maximum lag $n = n_a + n_b + n_c$ and on the order of the nonlinearity l, which determines the number of polynomial combinations between all input and output terms. In particular,

$$n_r = \frac{(n+l)!}{[n!l!]}.$$

For each model regressor, then, it is still necessary to estimate each single parameter value with the chosen optimization technique. These considerations make it difficult to apply polynomial techniques to the human–robot system, characterized by a high degree of nonlinearity and impulsive force.

6.1.2 *Neural network approximation*

The second strategy is to approximate the $f(\cdot)$ by using an artificial neural network (ANN). As stated in the universal approximation theorem, *"a neural network with a single hidden layer can accurately approximate any nonlinear continuous functional."* This consideration makes ANNs the obvious choice for our scope, even if more complex structures rather than the simple single-layer one indicated in the theorem are used nowadays (DeVore *et al.*, 2021). In mathematical terms, we have the following (Higgins, 2021).

Theorem 6.1. *Let $\phi(\cdot)$ be an arbitrary activation function. Let $X \subseteq \mathbb{R}$ and X be compact. The space of continuous functions on X is denoted by $C(X)$. Then, $\forall f \in C(X)$, $\forall \epsilon > 0 : \exists n \in \mathbb{N}$, a_{ij}, b_j, $w_i \in \mathbb{R}$, $i \in \{1 \ldots n\}$,*

$j \in \{1 \ldots m\}$:

$$(A_n f)(x_1, \ldots, x_m) = \sum_{i=1}^{n} w_i \phi \left(\sum_{j=1}^{m} a_{ij} x_j + b_j \right)$$

is an approximation of the function $f(\cdot)$; that is,

$$\| f - A_n f \| < \epsilon.$$

Therefore, the universality property is independent of the chosen activation function but is a consequence of the multilayer feedforward architecture, as proved by Hornik *et al.* (1991). This concept led researchers to extend such a property to different network structures to gain an advantage in modeling different kinds of systems and functionals, as in the case of the study by Schäfer and Zimmermann (2007), where the authors, starting from the work of Hornik, developed RNNs to map open-loop dynamical system. Shen *et al.* (2022) derived an approximation error bound with an explicit prefactor for Sobolev-regular functions using deep convolutional neural networks (CNNs), while in Csáji *et al.* (2001) improved the approximation performance by using a multiresolution approach which increases network robustness.

6.1.3 *Results*

Going back to our setup, the first step to do is the preparation of the dataset in order to correctly feed the network with the correct autoregressive, input, and error elements. The dataset used for training the network is the same as the one used in Chapter 3, previously used for linear model identification. All the signals have been processed considering x and y components and have been normalized between -1 and 1 before being fed to the network. The considered input signal was the randomly generated wave square, which was the reference that human subjects had to follow, while the output signal is, as mentioned, the generated force.

As schematically represented in Figure 6.1, in each time frame t, we ask the network to predict the signal $y(t)$ by finding the optimal internal connections (represented by the network weights values), starting from a window of input data which is composed of the following:

- $y(t-1), \ldots, y(t-n_a)$ force samples;
- $x(t-d), \ldots, x(t-d-n_b)$ reference signal samples;
- $e(t-1), \ldots, e(t-n_c)$ noise terms.

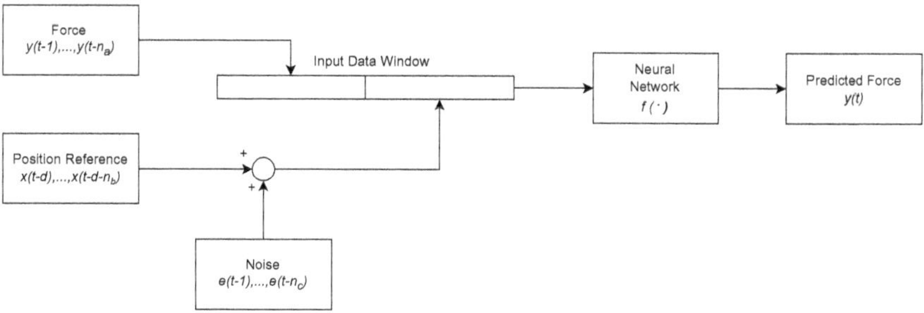

Fig. 6.1. Schematic representation of how the proposed NARMAX model was constructed by building an input data window at each time t and using an ANN to approximate the nonlinear functional element.

This window has to change for each sample of the force signal we want to predict and must be built in both the training, validation, and test sets. In our model, $n_a = n_b = n_c = 10$, so the window comprises 31 elements.

Noise terms, in particular, have an important role in system modeling. As seen in the linear case in Chapter 2, the human controller itself introduces noise into the system, to which the crossover model refers as *"remnant noise."* McRuer used to model such noise as a random process linearly uncorrelated with the control input. Physically, such a process is related to the error perceived by the human between the reference signal and perceived feedback and should be treated as an observation noise to add to the system input; the validity of this assumption was experimentally verified by McRuer and Krendel (1974). To stick with this important physical aspect of the human controller that we want to model, the noise terms $e(t - 1) \ldots e(t - n_c)$ and reference signal $x(t - d) \ldots x(t - d - n_b)$, which is our control input, were added. The resulting terms were used, along with the n_a force samples, as input to the network when building the data window illustrated above.

The final network structure that was chosen is a four-layer ANN with a rectified linear unit (ReLU) activation function, apart from the last layer, which was set as a linear layer as traditionally done so as not to restrict the output range of variation (Csáji *et al.*, 2001). The network was trained with a learning rate of 0.001 and a mean squared error loss function, which predicted most accurately each time instant of the $y(t)$ signal. The dataset was split into 50% used for training, 25% for validation, and 25% for testing. During the model training phase, the time instant of the force signal whose value has to be predicted has been randomly chosen within the range of indices between $[d + n_b, L_y]$, where L_y indicates the sample size of the signal. This process has to be repeated for each of the three aforementioned subsets, adapting L_y size according to the corresponding indicated percentages.

After the training stage has been completed, to test the model, the force values $y(t) \in [d + n_b, L_y]$ have been forecasted this time following a chronological order to correctly reconstruct the force signal during the whole duration of the experiment. Figure 6.2 shows the resulting de-normalized predicted force versus the measured one; the brown part represents the portion where the two signals are overlapped. The figure shows a summary of the dataset used for training, consisting of 200 elements, 100 for x and 100 for y components. The experiment was randomly chosen so that there is one for each subject for both components, so one each decade of the dataset. As evident from the plots, the model is capable of forecasting force samples with an outstanding degree of accuracy since the overlapped part is almost the entire amount of the two signals, despite the huge variability between the acquired signals that constitute the dataset in terms of noise and periodicity of the vectors.

To quantitatively evaluate the model's performance, root mean squared error (RMSE) and R^2 score were computed for each predicted signal from the 100 experiments that were used for collecting the dataset. As observed, the model's good performance is shown by the very low RMSE, which was preferred to normal mean squared error for its higher interpretability, being of the same unit of measure as the input data (in this case, Newton).

As evident from the data shown in Table 6.1, the RMSE, which, as mentioned, shows de-normalized data in Newton, is close to 0. The value of the R^2 score, which is a numeric value not constituted by a unit of measure and still considers normalized data, is close to its maximum value of 1. It is worthy of a remark on the fact that the R^2 score does not have, instead, a lower limit: a model with bad performance could have as low values as possible from its physical characteristics! Both indices testify to the really high performance obtained by the model and its validity.

A concern that may arise from the shown results is that the model overfits the dataset, being so accurate in the approximation of every generated force in all the noise conditions. If that is true, it would mean that the model is not generally applicable when varying human subjects and the other experimental conditions.

To answer this question, in the following section, we will test the model's generalization capabilities by applying it to a different dataset, with a different forcing function, and with a different robot with respect to what had been used here.

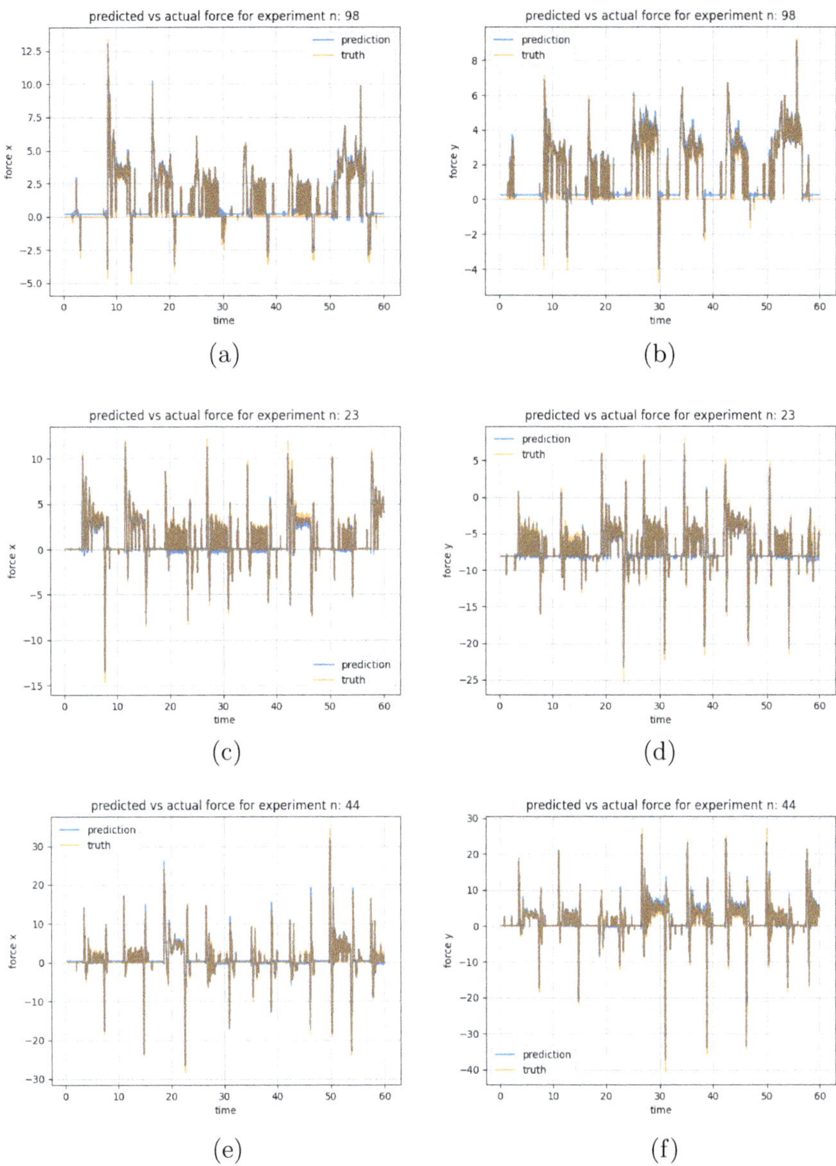

Fig. 6.2. Measured vs. predicted force values of both x and y components. Chosen experiments are a summary of 18 experiments extracted (1 every 10) from the original dataset of 100 experiments.

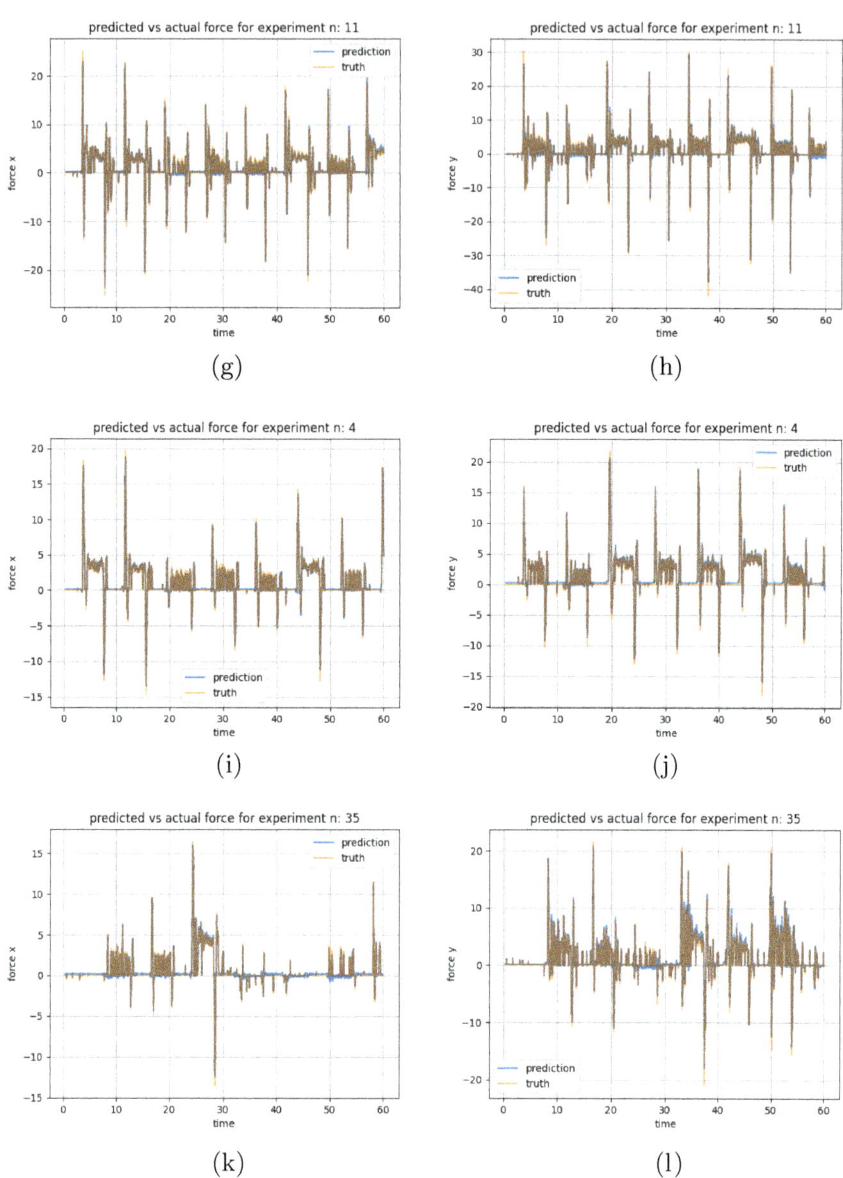

Fig. 6.2.   (*Continued*)

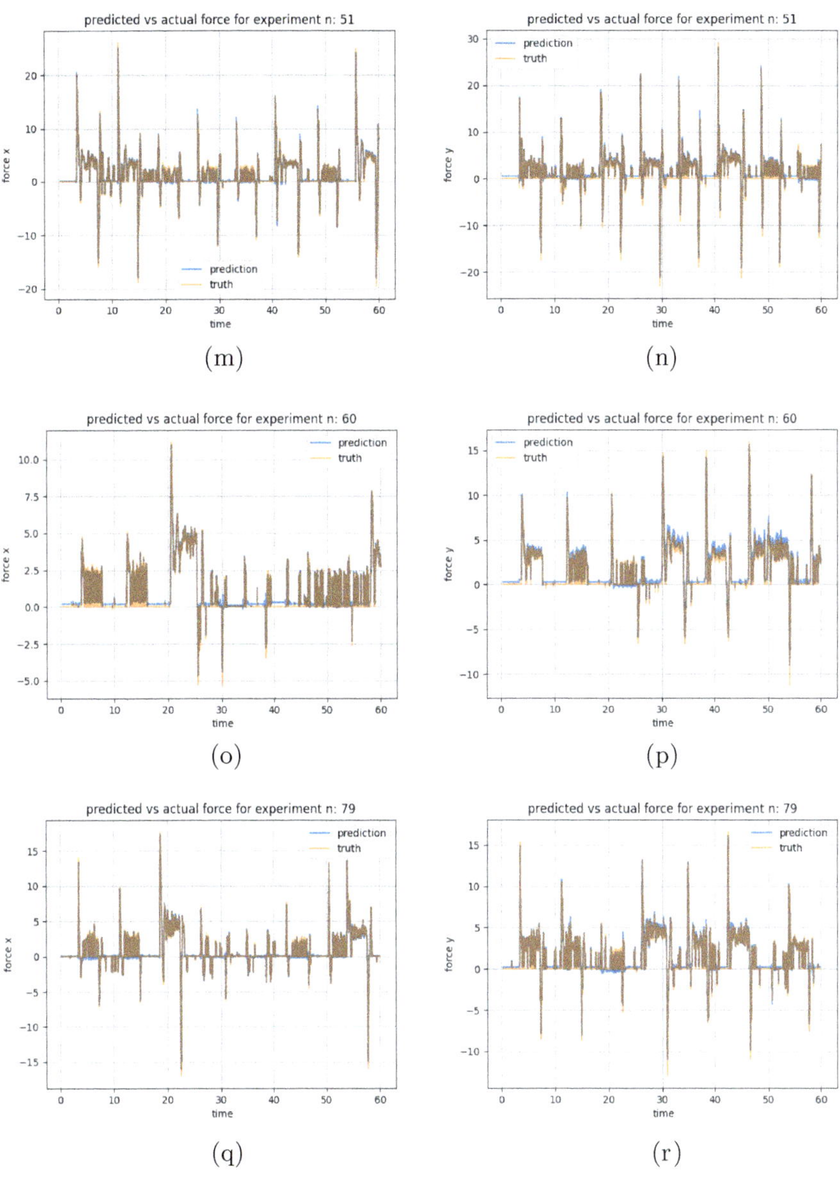

Fig. 6.2.   (*Continued*)

Table 6.1. Root mean square error (in N) and R^2 score considering the predicted versus the measured human force. The RMSE index, being expressed with the same unit of measure as the input data, considers de-normalized values. R^2 score is a numeric value with the upper bound of 1 and still uses normalized values.

Index	RMSE x	RMSE y	R2 score x	R2 score y
1	0.472036235	0.521664149	0.956724633	0.960830384
2	0.489941946	0.642806454	0.960653648	0.949716232
3	0.519541718	0.554902738	0.960189219	0.959170756
4	0.601814676	0.669788111	0.955643017	0.961106611
5	0.574182629	0.658225902	0.96087635	0.96252835
6	0.585448684	0.627132843	0.960482329	0.961794969
7	0.530537037	0.662696702	0.963967534	0.963649709
8	0.617680061	0.724884097	0.964767234	0.962247245
9	0.551281783	0.699222684	0.966718266	0.958319885
10	0.540838129	0.628847549	0.961557462	0.962917097
11	1.001928234	1.632743179	0.960195286	0.946946818
12	0.879766609	1.122209997	0.954856202	0.953832129
13	0.839163153	1.111029515	0.95914142	0.945890859
14	0.773373836	1.203074269	0.955129589	0.951072734
15	0.891532239	1.158637026	0.949615224	0.950984968
16	0.755985785	1.084383216	0.952572305	0.957884578
17	0.685293479	0.921445126	0.953228398	0.954294241
18	0.573845426	0.980308637	0.950972863	0.957289423
19	0.515555599	0.853679632	0.951384536	0.958947792
20	0.732953844	1.276626686	0.941025365	0.937022948
21	0.779908125	1.032217524	0.957220077	0.955973842
22	0.717266276	1.013400762	0.951455669	0.936464505
23	0.659451419	0.813392031	0.968066462	0.960385522
24	0.595282446	0.714604865	0.948058756	0.946671776
25	0.580491582	0.689179515	0.952205569	0.948194036
26	0.637227825	0.80138281	0.96229403	0.963255591
27	0.641417839	0.795405533	0.965880499	0.963530068
28	0.653279451	0.820438478	0.958745918	0.957266842
29	0.696505807	0.931492427	0.954515885	0.97026166
30	0.633404694	0.823071262	0.960112106	0.963946552
31	0.653850802	0.638911347	0.936389479	0.952612566
32	0.527259381	0.633462448	0.941242301	0.946880339
33	0.648364611	0.760147021	0.9448306	0.9525119
34	0.531802732	0.668122088	0.954120572	0.959334652
35	0.496948157	0.959522604	0.955091424	0.937008427
36	0.469225642	0.766567706	0.941583529	0.951641584
37	0.412226024	0.597905458	0.958859779	0.956849482
38	0.513679866	0.541526516	0.951304856	0.957271993
39	0.488406263	0.580186734	0.956285987	0.957061112
40	0.537667398	0.689487665	0.953796888	0.958131693

Table 6.1. (*Continued*)

Index	RMSE x	RMSE y	R2 score x	R2 score y
41	0.810000726	1.001200503	0.959480548	0.967295193
42	0.816242923	1.08466761	0.960760813	0.967652903
43	0.838387612	0.965088842	0.967119422	0.958562475
44	0.756781692	1.113235316	0.961284896	0.958956991
45	0.909424742	1.005840925	0.959332263	0.963989186
46	0.63409503	0.969677432	0.957599245	0.966572981
47	1.057900284	1.351292111	0.961301498	0.956206088
48	0.71832515	0.785362473	0.956490612	0.963244864
49	0.690414144	0.84771881	0.964459701	0.946220406
50	0.613183623	0.764978636	0.969409659	0.966009687
51	0.742447286	0.960810155	0.960201052	0.956050056
52	0.751854625	0.914295651	0.957280078	0.952400024
53	0.811494761	1.238216424	0.942534553	0.954774985
54	0.801926757	1.147018492	0.935714128	0.956892106
55	0.800856941	1.033023894	0.948548094	0.955067056
56	0.858715064	1.075729024	0.939555377	0.967784763
57	0.776043582	1.036302229	0.963472345	0.955151007
58	0.755968542	1.033446737	0.948050068	0.966379937
59	0.949616911	1.375363849	0.933001818	0.962863828
60	1.016029316	1.455973209	0.938462374	0.957735294
61	0.44272718	0.451795379	0.935553434	0.969999585
62	0.502331117	0.493711302	0.936199774	0.958587934
63	0.542155534	0.603473038	0.927183852	0.946375419
64	0.490480574	0.555177238	0.951736575	0.954753754
65	0.466246555	0.54974428	0.945880337	0.956059571
66	0.438220919	0.468733171	0.947413566	0.956953769
67	0.423949319	0.480391149	0.950661121	0.954565105
68	0.48883597	0.535027873	0.931848345	0.956498211
69	0.481457007	0.57267078	0.947277832	0.951858001
70	0.52182963	0.618782586	0.938644815	0.938375534
71	0.491386615	0.607104793	0.937202986	0.945512594
72	0.463411188	0.597224165	0.951821046	0.947026245
73	0.532284003	0.648179002	0.95797053	0.955360121
74	0.53142484	0.649036079	0.954641615	0.958479429
75	0.585957896	0.734377326	0.961540867	0.961957569
76	0.601249499	0.842698817	0.954111628	0.953575918
77	0.619303016	0.637448297	0.966617421	0.934785461
78	0.687656891	0.743184784	0.961381829	0.927876761
79	0.592401816	0.658418071	0.953752515	0.946422712
80	0.628193068	0.664767329	0.942279814	0.936807898
81	0.742072066	1.083717746	0.949598424	0.957108816
82	0.850080368	1.206421153	0.921889782	0.94543225
83	0.922698646	1.220767547	0.918371174	0.952035447
84	0.894367366	1.402564313	0.917593521	0.94120114
85	1.133547374	1.820029435	0.948960106	0.94123321

(*Continued*)

Table 6.1. (*Continued*)

Index	RMSE x	RMSE y	R2 score x	R2 score y
86	0.938934707	1.359940352	0.917720283	0.934916194
87	1.060884229	1.73831993	0.910323347	0.940331842
88	1.14246511	1.507356616	0.917072358	0.940996278
89	0.90528934	1.223504068	0.92624921	0.946063163
90	0.955415892	1.14667481	0.941171141	0.937761062
91	0.390547613	0.410679524	0.954216646	0.95810864
92	0.415624598	0.427568629	0.956806357	0.948533138
93	0.486593196	0.409988808	0.935246045	0.942162937
94	0.458686577	0.427382061	0.945552973	0.948642137
95	0.440944996	0.411954816	0.940034109	0.951950269
96	0.439976922	0.348811568	0.949425666	0.96367613
97	0.456322357	0.407657337	0.948594671	0.954207793
98	0.429119011	0.34507542	0.949096734	0.9616144
99	0.465654266	0.395928492	0.943786995	0.952985548
100	0.450937512	0.415887569	0.936107923	0.948991599

6.2   Human Intention Estimation

6.2.1   *Experimental setup*

The human model described in Section 6.1 was used to estimate human intention in a collaborative manipulation task, whose setup was similar to the one proposed in Chapter 3. The experimental setup has some differences with respect to what was proposed for the first time in the linear case and then used for building the nonlinear model in the previous section, where the model was built offline starting from the collected experimental data.

The model has now been tested online in a similar setup but with a different robot and different randomly generated reference signals.

In particular, a Comau NS16 robotic arm was chosen (Figure 6.3). This was done for two main reasons; the first is the weight difference: NS16 weighs 335 kg against the 20 kg of the UR5 robotic arm previously used. The difference in the mass of the robot is reflected in the inertia that the arm has during a motion; when a human operator is performing a manipulation task manually guiding the NS16 robot, suddenly stopping its movement when the target point is reached requires more effort than when UR5 is used. This aspect increases the importance of the proposed human intention estimation framework to help the robot anticipate its stop or start movement according to what the operator wants.

The second reason is also related to the added value of the model. The Comau robot, unlike the Universal Robot, is not built to be collaborative

Fig. 6.3.   Comau NS16 industrial robot.

but is an industrial robotic platform imagined to perform assembly tasks, lifting medium-sized payloads (its payload limit is 16 kg) in an isolated workspace. This means that, since the manufacturer did not give the robot collaborative capabilities in the first place, they can still be achieved, thanks to the proposed framework, by increasing its intelligence and adaptability (Scibilia *et al.*, 2024a).

Figure 6.4 represents the control logic structure of all the robotic platform's components. The robot controller communicates via the power line to the arm on one side and via the MQTT protocol with the PC client on the other. The PC client is equipped with the ROS Melodic software framework, which proved to be useful for sensor fusion and robot control with a high-level architecture in many applications and the ROS control library. In the ROS laptop, two software components were implemented to allow communication with the robot and its control with real-time performance:

- *robot hardware interface*, responsible for managing the communication between the robot and ROS control;
- *robot ROS controller*, where the proper high-level control logic is present.

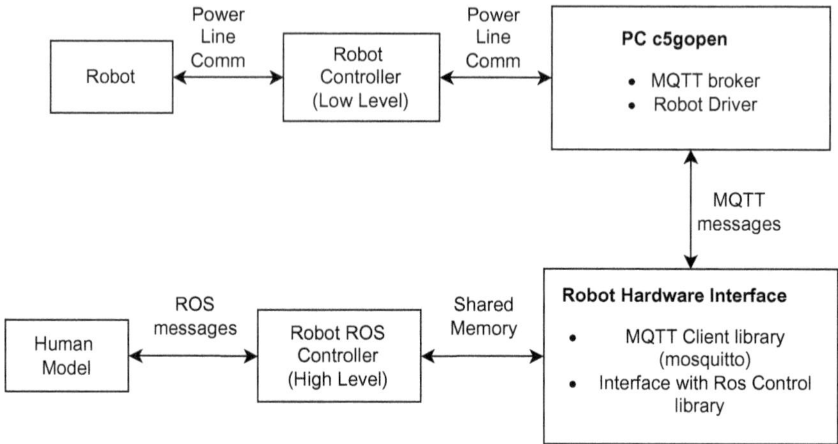

Fig. 6.4. Schematic representation of the hardware and software components constituting the robot control logic.

The robot hardware interface had the duty of managing outbound communication with the robot, implemented in the write method, and inbound communication, which gives robot feedback to the high-level controller, present in the read method. On one side, as mentioned, the robot communicates with the MQTT protocol; on the other, ROS control frameworks work with shared memory variables. For this reason, within the software component of the hardware interface, a C++ client MQTT library was implemented for converting the content of memory cells (message payload) into JSON messages, which were sent to the robot through the Mosquitto MQTT broker and vice versa.

The robot controller, similar to what was described in Chapter 3, contained the high-level admittance controller, which decided robot behavior during physical interaction with the human subject. Again, the Robotiq FT300 force-torque sensor was mounted on the robot end-effector to read the external force applied by the human. Once we know the force $\mathbf{F}_{\text{ext}}$, the position error $\Delta\mathbf{x}$, and the Cartesian velocity $\dot{\mathbf{x}}$ vectors, the acceleration commanded to the robot by the controller will be

$$\ddot{\mathbf{x}} = \mathbf{I}_{\text{pinv}}(\mathbf{F}_{\text{ext}} - \mathbf{K}\Delta\mathbf{x} - \mathbf{D}\dot{\mathbf{x}}), \tag{6.3}$$

where $\mathbf{I}_{\text{pinv}}$ represents the pseudo-inverse of the inertia matrix, while $\mathbf{K}$ and $\mathbf{D}$ are the stiffness and damping matrices, respectively. Since the Comau robot can be controlled in terms of position only, the computed acceleration $\ddot{\mathbf{x}}$ has to be integrated twice, before being sent.

To avoid numerical errors due to the excessive number of computations and numerical approximation of data, before being processed by equation (6.3), a deadband was applied to the external force vector, considering as null all the forces between -2 and 2 N. Moreover, for what concerns the velocity $\dot{\mathbf{x}}$, the following exponential filter was applied:

$$\dot{\mathbf{x}}_f = c\dot{\mathbf{x}} + (1 - c)\dot{\mathbf{x}}_f, \tag{6.4}$$

where c is a constant parameter that has been set at 0.8.

However, the admittance controller and the necessary data pre-processing are not the main peculiarities of this work. The main difference with respect to the previously proposed controller consists of an additional software component that communicates with the robot controller: the Python script containing the human model predictor.

The human model described in the first section has been loaded into a ROS node using the Pickle Python library. First, the force and the actual position vectors have been acquired from the FT sensor and the robot, respectively. Both signals have been stored in queues of a fixed size of 50 samples, a trade-off between having queues as short as possible and having a safety margin of at least 10 samples in case of failure from one of the two considered sources of information. Secondly, both queues have been normalized between -1 and 1, necessary to be processed by the model, which has been trained with normalized values. Then both queues were used to build the input data window, as explained in the first section, and the model forecasted the next x and y force components. Finally, the predicted values have been de-normalized and sent to the main admittance controller.

On its side, the controller needs a position value to consider as the next reference position and anticipate human motion. For this reason, before being used, the predicted human force was converted into velocity by multiplying its value by a factor of $\frac{1}{d_v}$, which corresponds to the reciprocal of a virtual damping constant d_v, and then integrated into the final position value.

Even though the application scenario in which this work is focused is the continuous physical interaction and a proper trajectory planner is not needed, a state-machine motion planning stage was necessary to avoid unstable and dangerous behavior:

- **State 0**: The current reference is the starting position, and the motion is disabled. A new reference is received, and the controller checks if such

reference is between an interval comprised between 1 mm and 0.3 m. If such a condition is verified, the motion is enabled.
- **State 1**: Motion is enabled; the robot's goal is to minimize the position error between actual and reference positions with a speed depending on the stiffness value set to the controller.
- **State 2**: The new goal position has been reached, the current position is set as the new starting position, and motion is disabled.

As said, the experimental setup was similar to the one described in Chapter 2, with a different reference signal for each experiment. Again, the reference signal was generated as a sequence of points randomly chosen with x and y components in the range of $[-0.1, 0.1]$ m from the starting position, which was set as a reference point in the sequence between every two random points. This way a complete step is performed for every generated point; the resulting forcing reference function is a square wave consisting of a sequence of steps.

Human subjects were asked to manually guide the robot toward the next reference position, displayed as a yellow dot on a screen showing the x–y plane also containing the robot model. The resulting experimental setup is shown in Figure 6.5.

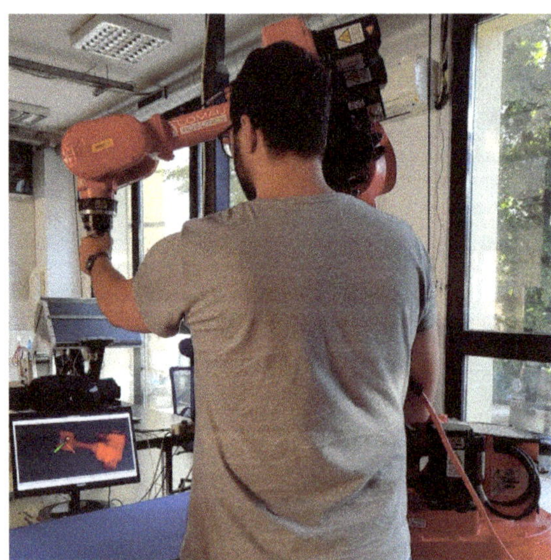

Fig. 6.5.   Experimental setup with a human subject performing a manual guidance task with the robot, following the virtual reference displayed on a screen.

The model, as mentioned, was used to forecast the next human force online. For this reason, rapid computational time is required, so both the human estimator node and the ROS robot controller node (using the same nomenclature indicated in Figure 6.4) were synchronized to run with a loop period of 0.00125 s, which corresponds with good approximation to 800 Hz.

In some experiments, the state machine, which converted the predicted force to a position reference for the robot, was used; in others, it was not.

6.2.2 *Results*

The obtained results in the case in which only the admittance controller was used are shown in Figure 6.6, while Figure 6.7 shows the results observed when the forecasted force was converted first to a velocity and then to a position reference, which was sent to the robot using the state-machine approach described above. Below each force plot is also shown the correspondent position tracking error, one for each component.

In all the plots, the first rows, showing two superimposed vectors, represent the force's x and y components, respectively. The blue curve represents the measured force acquired by the FT300 force sensor, while the orange curve is the force forecasted by the model.

The first observable aspect is that all the force curves present an initial temporal shift, which in some cases is of 3–5 s. Going forward with the experiment, the time delay between the two force signals kept decreasing until it became negligible in the last 10 s of the experiments. This happens due to the fact that the two queues used as input windows from the model were initialized with null values in all their elements at the beginning of each experiment. This caused a "temporal offset" since the model needed some time to fill its two queues of 50 samples each with the real measured values. This requires 0.0625 s only to fill the queues and more time to reflect this change in the model output, which is obviously zero at the beginning.

Despite this, the signal produced by the model accurately reflects the measured force even in the first "tuning part" of the experiments, where the time delay was considerable and remained high until the last part, in which the two signals are almost superimposed.

Another consideration is that in the majority of cases, the robot ROS controller failed to distinguish between unwanted external contact, which must be managed in a compliant way according to the admittance control law of equation (6.3), and force produced by the human subject, which must be converted into a position reference as quickly as possible. The state-machine algorithm and the model alone appear insufficient to perform such a complex task.

However, when the human reference was correctly recognized, such as in Figure 6.7(a–d) and in parts of Figure 6.7(i–l), the advantages in terms of performance are evident. In fact, the force signals present only the first peak, after which the robot reaches the position desired by the human operator; the successive noisy part of the force signal, which was observable in the previously used data, is not present anymore.

Also, the position tracking error indicates a visible improvement, decreasing from over 0.1 m present in other cases to the maximum of 0.42 m observable in Figure 6.7(c) and 0.29 m in Figure 6.7(d). Even the central part of Figures 6.7(k) and 6.7(l), where the characteristic impulsive behavior is evident, confirms this consideration.

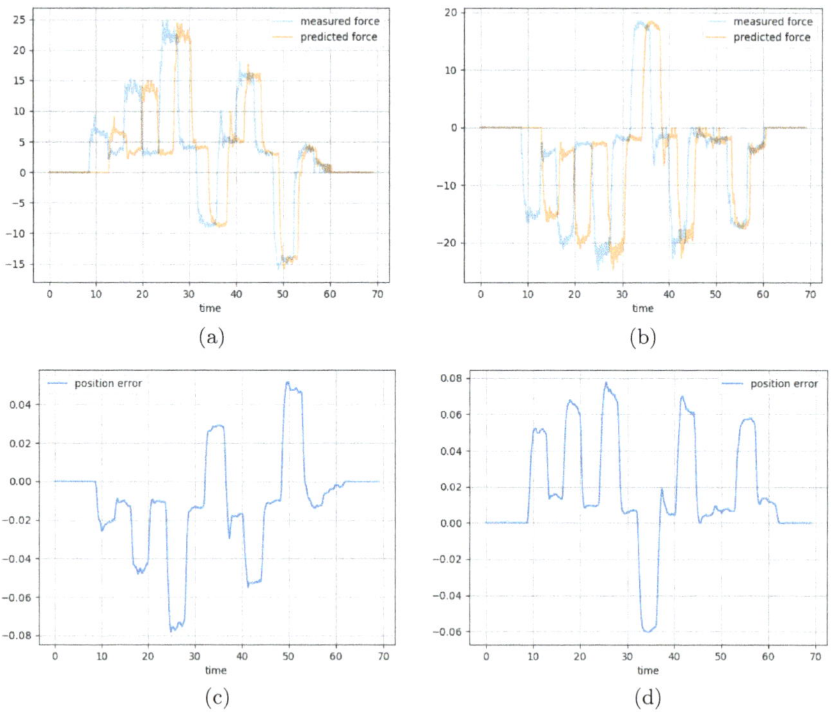

Fig. 6.6. Measured vs. predicted force values (first rows) and position tracking error (second rows) of both x (first columns) and y (second columns) components for each experiment. Differently from what was previously done, force value is now forecasted online.

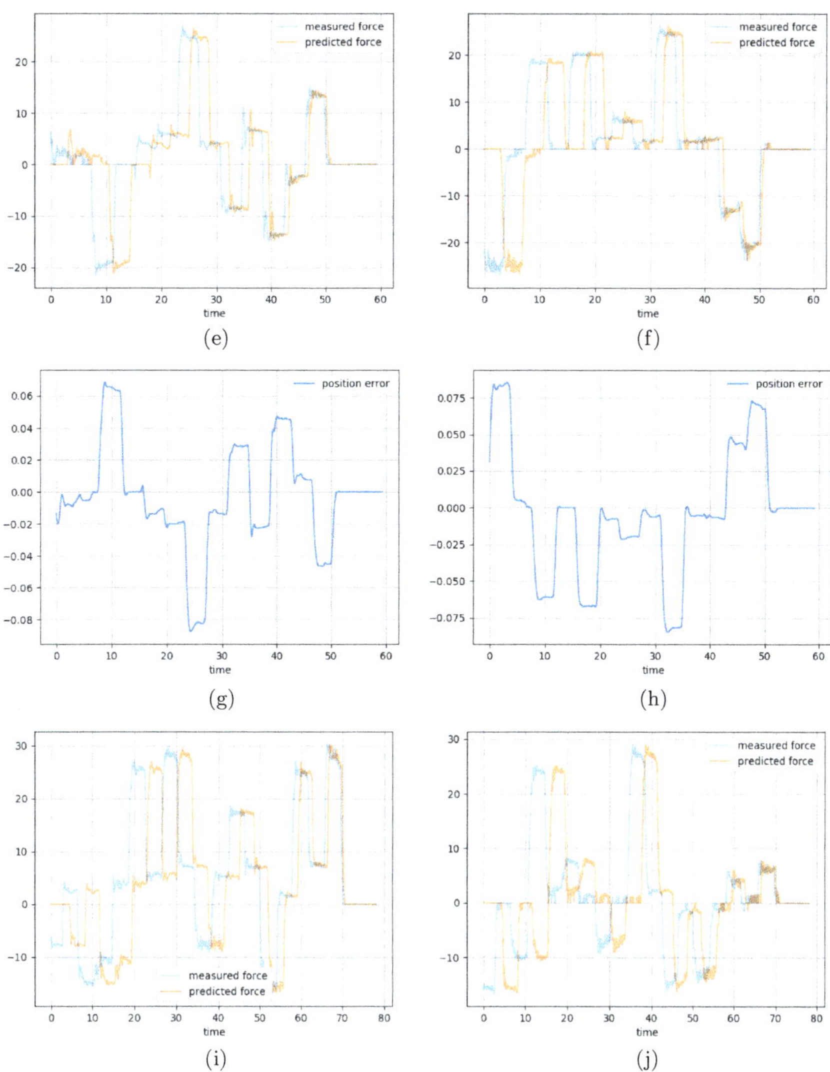

Fig. 6.6. (*Continued*)

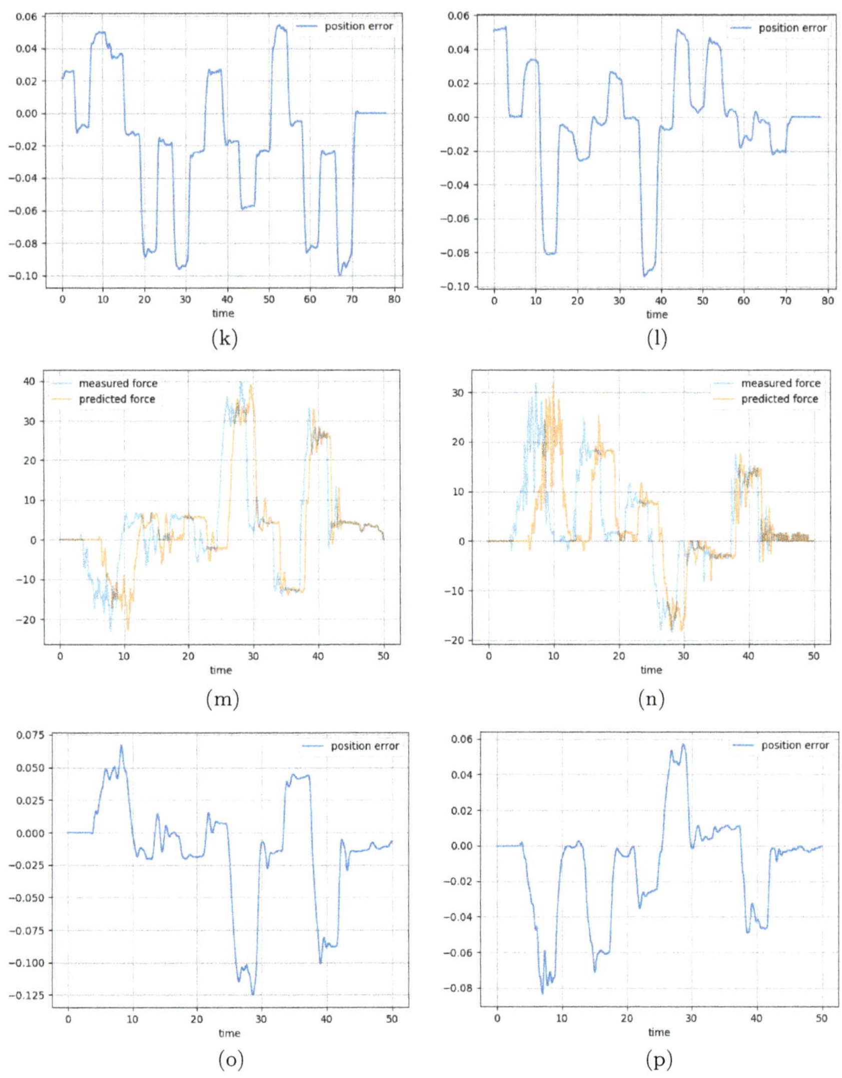

Fig. 6.6. (*Continued*)

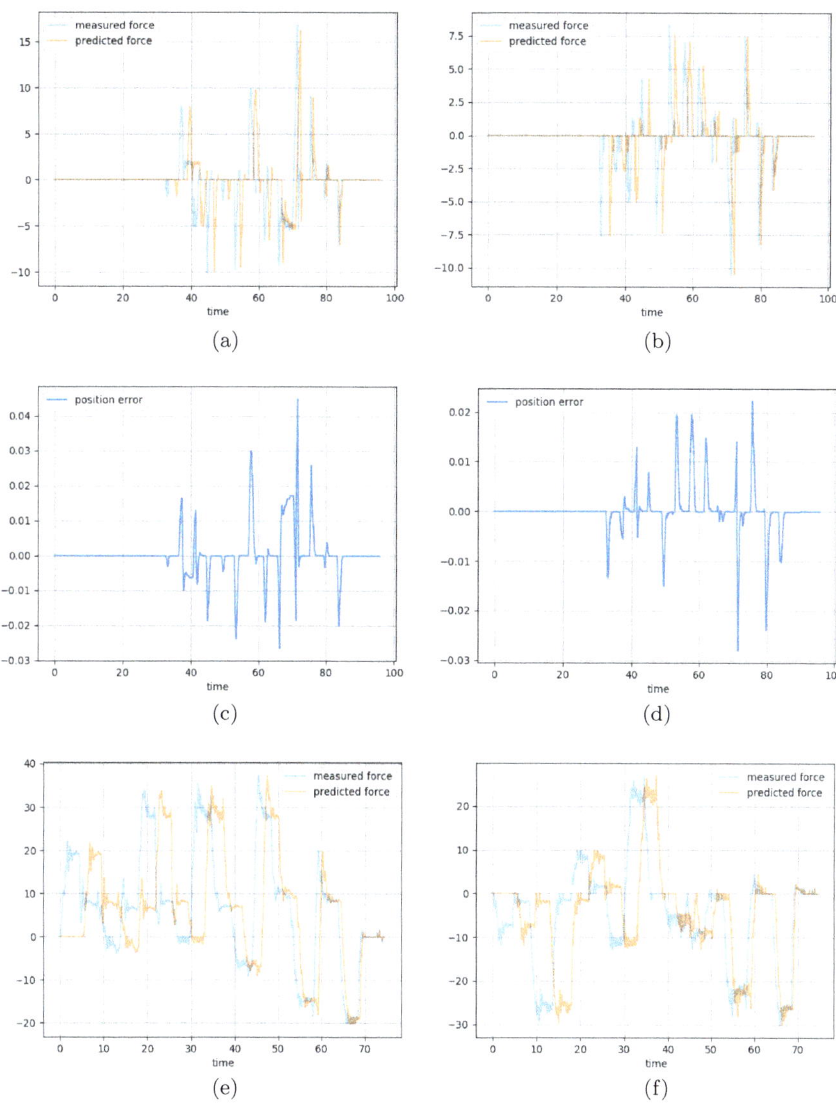

Fig. 6.7. Measured vs. predicted force values (first rows) and position tracking error (second rows) of both x (first columns) and y components (second columns) for each experiment using the state-machine approach to convert it into a position reference.

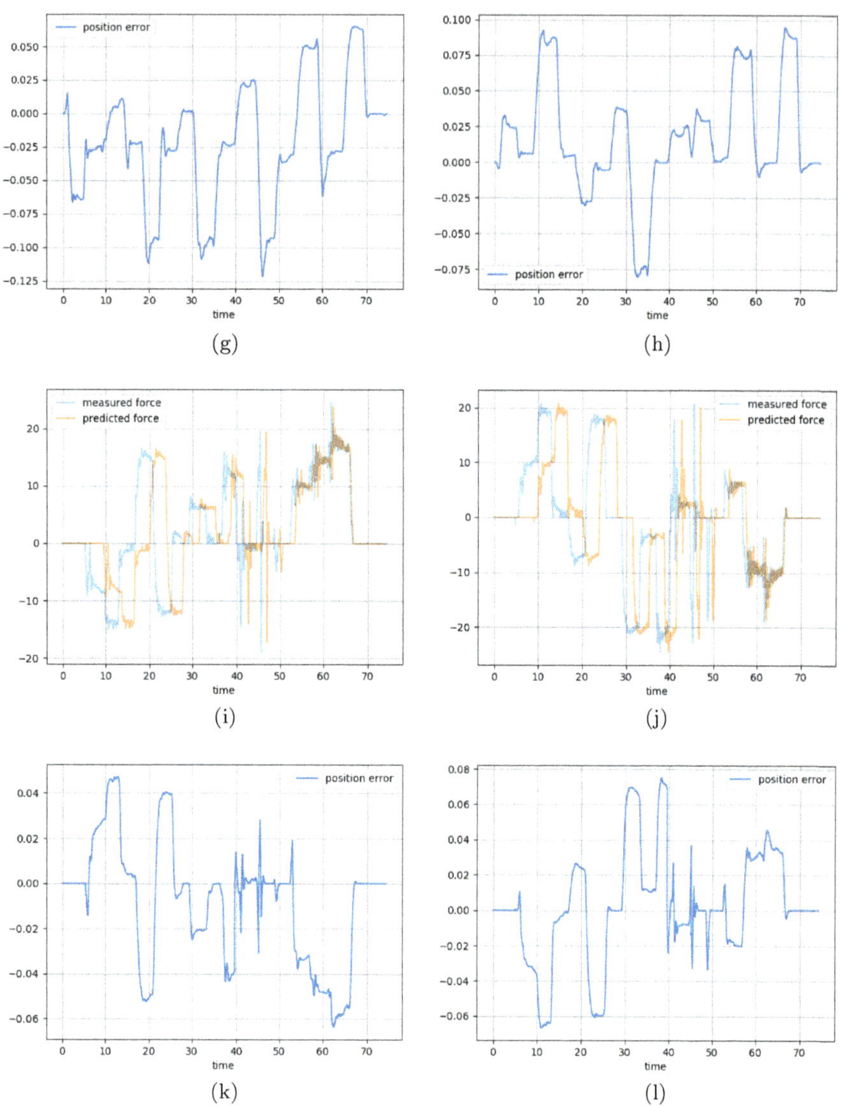

Fig. 6.7.   (*Continued*)

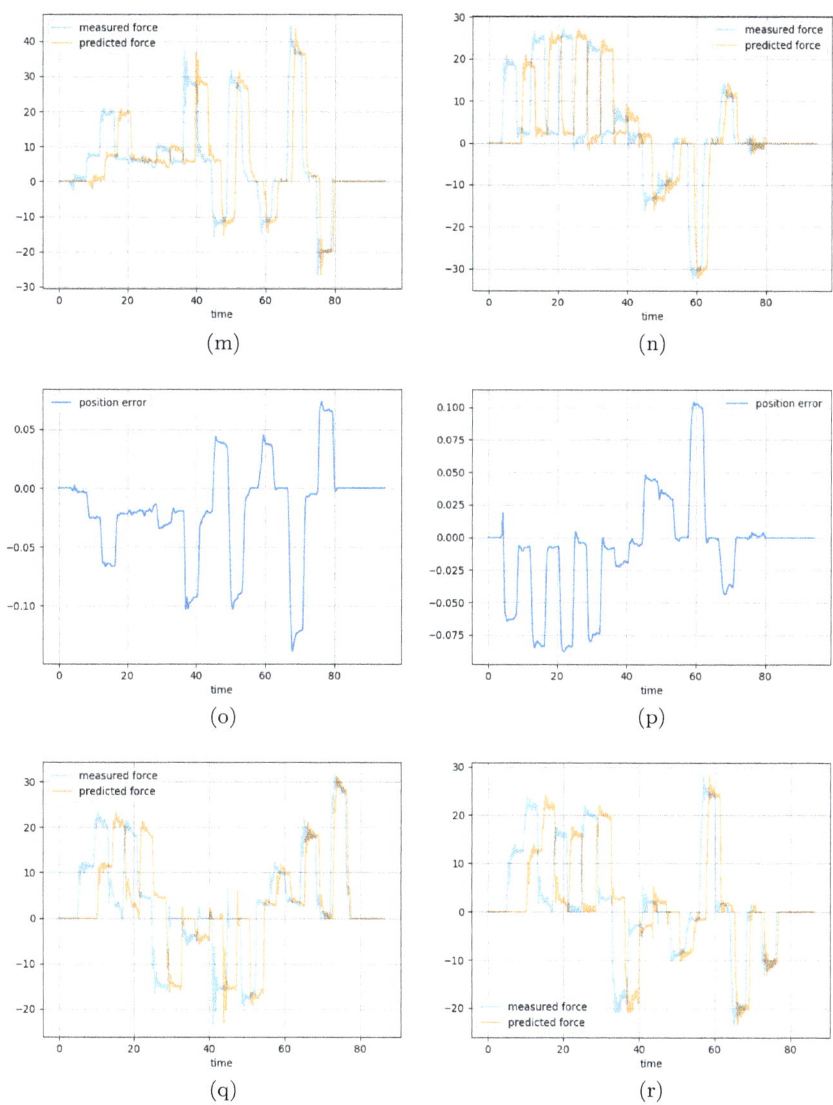

Fig. 6.7.　(*Continued*)

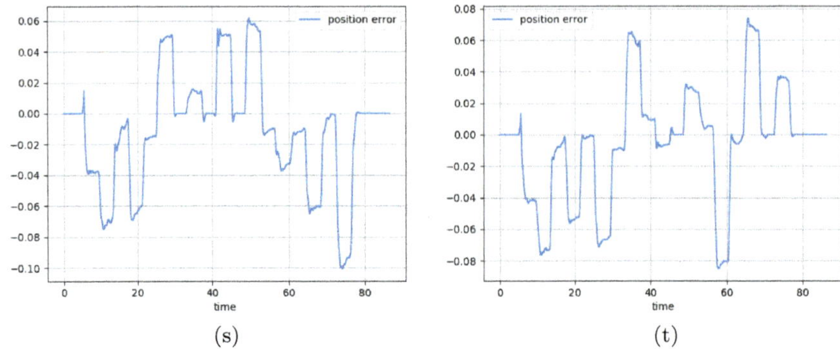

(s)   (t)

Fig. 6.7. (*Continued*)

Chapter 7

Frequency Domain Analysis and Human Reactive Delay Estimation

The simple linear model proposed in Chapter 2 provided a clear description of the system's frequency response and was able to explicitly extract the estimation of human gain and delay, even if only after some additional data processing.

As mentioned in Section 6.1, complex impulsive and highly nonlinear dynamics observable in human–robot interaction struggle to be represented entirely in all its aspects by techniques assuming to approximate the system to known simple linear equivalents, such as in the case of the crossover model experimentally studied in detail in Section 3.2. For this reason, the use of a more complex technique, such as Narmax models, which are able to deal with unknown dynamics, helped to increase the level of the model's accuracy.

However, until now, we have limited our analysis of human control actions during robotic manipulation tasks in terms of time series. In this chapter, we try to extend the way in which our model discussed in the last chapter can be used, moving our focus on the frequency domain to explicitly provide all the human parameters of interest directly.

7.1 Introduction

The accurate estimation of control delays is critical for advancing human–robot interaction, where timely and precise responses ensure safety, efficiency, and collaboration. Over the years, various approaches have been proposed, ranging from linear models to advanced nonlinear techniques

117

like the nonlinear autoregressive moving average with exogenous inputs (NARMAX) model. These methods have been explored extensively in tasks requiring dynamic adaptation and precision.

Linear models have traditionally dominated the field due to their simplicity and computational efficiency. However, their inherent limitations in capturing complex, nonlinear dynamics often make them inadequate for real-world human–robot interaction scenarios. For instance, linear models assume fixed relationships between inputs and outputs, failing to account for the adaptive and time-varying behavior exhibited by humans during interaction tasks. Such constraints necessitate the exploration of nonlinear alternatives.

The NARMAX model has emerged as a powerful tool in this domain, offering the flexibility to model dynamic and nonlinear systems effectively. Research by Billings and Coca (2008) has highlighted the capability of NARMAX to capture intricate temporal dependencies, making it particularly suitable for estimating control delays. These models extend traditional linear approaches by incorporating nonlinear terms and memory effects, enabling a richer representation of the underlying dynamics. In tasks like robotic manipulation and trajectory prediction, NARMAX has been shown to outperform linear models, especially in environments where human behavior introduces significant variability.

Comparative studies have further underscored the superiority of NAR-MAX over traditional methods in the estimation of human control parameters. For example, Akanyeti *et al.* (2008) explored scenarios where linear models failed to adapt to the nonlinearities inherent in human–robot interaction. The study revealed that NARMAX could account for these complexities, providing a more accurate representation of system behavior. Moreover, the incorporation of hybrid approaches — combining NARMAX with data-driven techniques — has pushed the boundaries of what these models can achieve. Research by Zhu *et al.* (2022) has shown that hybrid NARMAX models, when integrated with machine learning algorithms, can achieve unparalleled accuracy in estimating control delays, even in highly dynamic settings.

Overall, the evolution from linear models to advanced nonlinear techniques reflects the growing need for robust and adaptive frameworks in human–robot interaction. The adoption of NARMAX represents a significant step forward, addressing the limitations of earlier approaches and paving the way for more responsive and intelligent systems. By bridging the gap between theory and application, these models have set a new standard for control delay estimation, ensuring that robots can operate seamlessly alongside humans in diverse environments.

However, in cases where the system is partially known, its complexity can be reduced with different strategies. In Chapter 12, the use of peak-to-peak dynamics for this scope will be explored, while here we will see how the knowledge of part of the system can help extract from raw experimental data useful information which instead would require more complicated and time-demanding techniques.

As we saw in Chapter 3, the estimation of time delay, considering different subjects, can require a certain effort. In the proposed modeling technique, in fact, it was necessary to perform preliminary data processing to obtain the derivative of the position error useful to approximate the reaction time heuristically; then the use of identification techniques to optimize the model's parameters so that it would be able to simulate the human–machine system's response.

Now, having a reliable model of the human response only, producing an output force signal, which is the input of the robot control law expressed in equation (6.3), the knowledge of the controlled element's dynamics will allow us to obtain all the system's parameters in a fast and equally reliable way.

7.2 Methodology

In order to better investigate the control delay of the model in the context of human–robot interaction, the transport delay was considered as the time lag between the instant in which the human subject receives the stimulus from the forcing function and generates a force and the instant in which the robot receives such force. So, if we consider a simple block scheme of the system in which human and robot elements are sequentially connected, such as in the case of the linear CO model of Chapter 2, we ideally are now between the two blocks.

First of all, we have to estimate the robot transfer function. In our collaborative manipulation task, robot dynamics can be approximated as a second-order transfer function with a gain k, a natural frequency ω_n, and damping ratio ϵ:

$$F(s) = \frac{k}{s^2 + 2\epsilon\omega_n s + \omega_n^2}. \tag{7.1}$$

As shown in Figure 7.1, it is necessary to extract the frequency and damping parameter values from the human output $y(s)$ to build the robot

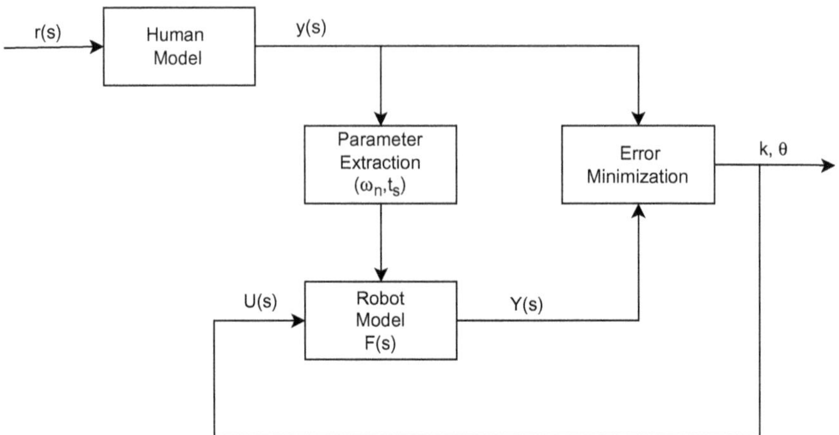

Fig. 7.1.   Schematic representation of the delay estimation algorithm.

transfer function. To do so, the force signals have been segmented into sections of 5 s duration each. For each window of signal, the oscillation period T was identified as the ratio between the window duration and number of oscillations, then

$$\omega_n = 2\frac{\pi}{T}.$$

Knowing the natural frequency, the ϵ parameter was derived from the settling time formula:

$$t_s = \frac{3.9}{\epsilon\omega_n},$$

where $\tau = \frac{1}{\epsilon\omega_n}$ in the system time constant and t_s is the settling time approximated after 3.9τ.

Once the transfer function F(s) of the robot controller is built, its output signal $Y(s)$ can be expressed as $Y(s) = F(s)U(s)$, where the input signal $U(s)$ is

$$U(s) = \frac{1}{s} - \frac{1}{s}e^{-s\theta}. \tag{7.2}$$

Here, θ represents the delay itself. A fourth-order Padé polynomial can well approximate the exponential part of the function. This way, being the input of the robot function directly calculated from the delay θ, and the robot function itself dependent only on the gain k, the optimal k and θ parameters can be estimated by an iterative algorithm, minimizing the root mean squared error between the simulated $Y(s)$ signal and the output of the human model $y(s)$. This final stage is represented in Figure 7.1 by the error minimization block and the subsequent loop. In other words, this process is equivalent to anticipating the robot model's output until it coincides with an acceptable degree of accuracy with the output of the human model.

The results are shown in Table 7.1. It can be noted that the oscillations of optimal parameter value between each experiment are almost negligible for the gain k and small for the time delay θ, which is distributed around the value of 0.3 s.

To estimate the control delay using the NARMAX model, we summarize the systematic approach that we used from the perspective of delay identification, as depicted in Figure 7.1:

(1) **Data collection**: The dataset used was done with the same experiments reported in the linear case with the UR5 robot.
(2) **Human model prediction**: The NARMAX model simulates and predicts force data, generating a time series.
(3) **Parameter estimation**: The frequency characteristic of the system response is extracted by the parameter estimation block, which computes the natural frequency and the settling time of the system given time window of 5 s.
(4) **Robot model prediction**: The force output of the robot model is simulated, given the robot controller's equation and the estimated parameters.
(5) **Window-shifting**: The error between the two responses is minimized by modulating the gain of the robot's simulated force output and time-shifting it according to a certain factor, which, in our case, simulates reaction delay.

This process doesn't only replicate the functionality achieved by the linear precision model for what concerns the estimation of reactive control delay but overcomes them by achieving a more precise characterization of the system's frequency response, enhancing the added advantages of nonlinear modeling and their capability of capturing intricate dynamics.

7.3 Results and Discussion

The results obtained using the NARMAX-based nonlinear modeling approach for delay estimation demonstrate significant advancements

Table 7.1.   Identified parameters for the controlled element's transfer function and delay estimation.

Index	Period	Epsilon	Omega	Gain	Delay	RMSE
1	0.1555	0.0193	52.181	2.5	0.298	2.2475
2	0.1296	0.0161	56.634	2.5	0.29	2.4750
3	0.1334	0.0166	53.437	2.5	0.278	2.4905
4	0.1548	0.0192	51.724	2.4	0.296	3.1595
5	0.1517	0.0188	51.953	2.5	0.267	3.0525
6	0.1546	0.0192	50.468	2.5	0.287	2.9547
7	0.1658	0.0206	49.669	2.4	0.286	2.9653
8	0.1636	0.0203	52.866	2.5	0.29	3.3834
9	0.1519	0.0189	51.839	2.4	0.281	3.2193
10	0.1551	0.0193	53.209	2.5	0.283	2.9257
11	0.1450	0.0180	52.181	2.6	0.325	5.3826
12	0.1217	0.0151	57.662	2.5	0.327	4.6006
13	0.1194	0.0148	57.433	2.5	0.334	4.3815
14	0.1155	0.0143	60.402	2.5	0.326	4.5168
15	0.1182	0.0147	58.004	2.5	0.326	4.4452
16	0.1098	0.0136	60.174	2.5	0.326	3.9863
17	0.1268	0.0158	54.921	2.5	0.338	3.5094
18	0.1253	0.0156	54.579	2.5	0.327	3.6054
19	0.1046	0.0130	62.229	2.5	0.306	3.2054
20	0.2161	0.0268	56.977	2.5	0.321	3.8185
21	0.1010	0.0125	66.111	2.7	0.335	3.5022
22	0.1026	0.0127	64.513	2.6	0.326	3.1613
23	0.1028	0.0128	64.513	2.6	0.327	3.4248
24	0.1106	0.0137	61.087	2.5	0.315	2.6977
25	0.1262	0.0157	58.575	2.5	0.321	2.6062
26	0.1054	0.0131	63.599	2.5	0.32	3.4739
27	0.1202	0.0149	59.032	2.5	0.313	3.7603
28	0.1154	0.0143	58.804	2.5	0.299	3.6272
29	0.1129	0.0140	60.174	2.5	0.324	4.1181
30	0.1161	0.0144	58.804	2.5	0.296	3.5908
31	0.1232	0.0153	57.890	2.6	0.315	2.5276
32	0.1428	0.0177	53.780	2.5	0.308	2.2940
33	0.1294	0.0161	57.662	2.5	0.3	2.9136
34	0.1425	0.0177	51.839	2.5	0.321	2.8352
35	0.1504	0.0187	50.240	2.5	0.297	3.0548
36	0.1676	0.0208	53.437	2.5	0.336	2.7918
37	0.1377	0.0171	54.465	2.5	0.311	2.3276
38	0.1938	0.0241	50.925	2.5	0.301	2.1996
39	0.2804	0.0348	52.524	2.5	0.32	2.3290
40	0.1643	0.0204	50.583	2.5	0.296	2.9647
41	0.1232	0.0153	55.378	2.5	0.352	4.2927
42	0.1188	0.0148	56.177	2.5	0.308	4.4177

Table 7.1. (*Continued*)

Index	Period	Epsilon	Omega	Gain	Delay	RMSE
43	0.1214	0.0151	55.721	2.5	0.332	4.5774
44	0.1270	0.0158	56.063	2.5	0.314	4.5013
45	0.1338	0.0166	51.153	2.5	0.306	4.8120
46	0.1249	0.0155	55.607	2.6	0.311	3.9243
47	0.1307	0.0162	53.551	2.5	0.321	5.8130
48	0.1181	0.0147	57.548	2.5	0.312	3.2490
49	0.1455	0.0181	49.441	2.5	0.329	3.8370
50	0.1197	0.0149	57.433	2.5	0.305	3.5187
51	0.1177	0.0146	59.375	2.5	0.309	3.8801
52	0.1200	0.0149	57.890	2.5	0.314	3.6323
53	0.1228	0.0153	55.721	2.5	0.311	4.4842
54	0.1213	0.0151	58.918	2.5	0.313	4.2957
55	0.3149	0.0391	49.212	2.4	0.338	4.2891
56	0.1286	0.0160	54.807	2.5	0.32	4.5250
57	0.1211	0.0150	57.890	2.5	0.312	4.1196
58	0.1240	0.0154	53.894	2.5	0.323	4.0612
59	0.1312	0.0163	53.437	2.5	0.313	5.2791
60	0.1283	0.0159	58.918	2.5	0.311	5.4367
61	0.1356	0.0168	55.264	2.6	0.307	1.8006
62	0.1333	0.0166	56.292	2.7	0.309	1.7296
63	0.1204	0.0150	59.603	2.6	0.312	1.9298
64	0.1324	0.0165	55.264	2.6	0.298	2.0855
65	0.1266	0.0157	54.465	2.6	0.316	1.9340
66	0.1326	0.0165	56.063	2.6	0.314	1.7050
67	0.1574	0.0195	49.212	2.6	0.304	1.8004
68	0.1215	0.0151	57.890	2.6	0.31	1.9447
69	0.1150	0.0143	62.115	2.6	0.31	1.9713
70	0.1224	0.0152	56.977	2.6	0.304	1.9737
71	0.1295	0.0161	54.465	2.6	0.317	2.1282
72	0.1371	0.0170	52.980	2.6	0.323	2.0923
73	0.1326	0.0165	55.264	2.5	0.321	2.5884
74	0.1397	0.0174	53.894	2.5	0.311	2.6593
75	0.1443	0.0179	52.295	2.5	0.307	3.2107
76	0.1227	0.0152	56.748	2.5	0.295	3.3325
77	0.1312	0.0163	52.866	2.5	0.324	2.8992
78	0.1250	0.0155	56.977	2.5	0.322	2.9160
79	0.1382	0.0172	50.240	2.5	0.308	2.7119
80	0.1271	0.0158	55.835	2.5	0.312	2.4921
81	0.3345	0.0415	52.752	2.4	0.29	4.1356
82	0.1163	0.0144	59.831	2.4	0.293	4.1216
83	0.1165	0.0145	59.260	2.4	0.292	4.4143
84	0.1121	0.0139	61.430	2.4	0.29	4.4727
85	0.1109	0.0138	62.115	2.5	0.282	5.8097

(*Continued*)

Table 7.1. (*Continued*)

Index	Period	Epsilon	Omega	Gain	Delay	RMSE
86	0.1116	0.0139	58.918	2.4	0.296	4.3111
87	0.1170	0.0145	58.233	2.3	0.275	5.9275
88	0.1149	0.0143	62.914	2.4	0.267	5.3416
89	0.1123	0.0140	61.658	2.4	0.281	4.5463
90	0.1169	0.0145	60.288	2.4	0.282	4.4289
91	0.1315	0.0163	58.918	2.7	0.33	1.4029
92	0.1324	0.0164	56.177	2.7	0.33	1.5262
93	0.1239	0.0154	58.119	2.7	0.341	1.3392
94	0.1267	0.0157	54.807	2.7	0.32	1.4988
95	0.1134	0.0141	60.060	2.7	0.328	1.3432
96	0.1333	0.0166	55.721	2.7	0.327	1.3807
97	0.1380	0.0171	51.839	2.7	0.329	1.4723
98	0.1595	0.0198	49.327	2.7	0.326	1.3727
99	0.1752	0.0218	48.756	2.7	0.315	1.4545
100	0.1438	0.0179	50.925	2.7	0.322	1.4086

compared to the linear precision model discussed in Chapter 3. While the linear model estimated delays between 0.4 s and 0.6 s (as shown in Table 3.1 of Chapter 3), the NARMAX model consistently identified delays in the range of 0.2–0.3 s. This marked reduction in delay values aligns better with established physiological benchmarks for human visual-motor reaction times, underscoring the enhanced accuracy of the nonlinear approach.

The physiological plausibility of the NARMAX-derived delays is supported by extensive research on human reaction times to visual stimuli. Studies have consistently reported reaction times within the range of 200–300 ms. For instance, Mangun and Hillyard (1991) observed reaction times between 200 and 250 ms during visual-spatial tasks, while Arruda *et al.* (1999) found a similar range (200–300 ms) in studies incorporating P300 responses. Other works, such as Fischer and Weber (1997) and Prinzmetal *et al.* (2005), reaffirm these findings, with reaction times reported between 220 and 300 ms, depending on task complexity and environmental conditions. Table 7.2 summarizes these physiological benchmarks, which further validate the improved accuracy of the NARMAX model.

The discrepancy between the results of the two approaches highlights the limitations of the linear precision model, which oversimplifies the dynamics of human–robot interaction by assuming static, time-invariant system behavior. These assumptions often lead to overestimations in delay,

Table 7.2. Summary on the state-of-art findings on physiological reaction times for visual stimuli over different scenarios and in different contexts.

Study	Findings	Reported range (ms)
Mangun and Hillyard (1991)	Modulations in perceptual processing during visual-spatial tasks	200–250
Fischer and Weber (1997)	Reaction times to visual stimuli under various conditions	220–300
Arruda *et al.* (1999)	Effects of exercise and gender on visual P300 responses	200–300
Bentin and McCarthy (1994)	Immediate repetition effects on reaction time in complex tasks	250–300
Roth *et al.* (1984)	Physiological responses to auditory and visual noise bursts	210–270
Andreassi (2010)	General human behavioral response times to visual stimuli	200–300
Prinzmetal *et al.* (2005)	Mechanisms of attention influencing reaction times	200–280
Langner *et al.* (2010)	Mental fatigue effects on simple reaction time tasks	220–290

particularly in tasks characterized by complex, time-varying inputs. In contrast, the NARMAX model captures intricate nonlinear dynamics and system memory effects, allowing for a more nuanced understanding of the temporal relationships between human actions and robotic responses.

Furthermore, the proposed approach demonstrates robustness against variability and noise in experimental data. In the linear case, individual subjects exhibited broader ranges of delay variation, as shown in Figure 3.4 of Chapter 3. By contrast, as noticeable by looking at all the parameters over the dataset experiment in Table 7.1, the NARMAX model produced narrower delay distributions, reflecting greater consistency across trials and subjects. This consistency is critical for practical applications, where reliable delay estimation directly impacts the performance of collaborative tasks.

In addition to delay estimation, the NARMAX-based model provides deeper insights into task-specific adaptations. By accurately modeling the frequency characteristics of human control actions, the approach enables the design of more adaptive and responsive robotic systems. These findings are particularly relevant in scenarios requiring real-time interaction, where even small improvements in reaction time can significantly enhance safety and efficiency.

7.4 Final Considerations

This chapter has demonstrated the significant advancements achieved by transitioning from linear to nonlinear modeling approaches for human control delay estimation in collaborative human–robot systems. By leveraging knowledge of the robot's transfer function, the delay estimation process benefits from a comprehensive understanding of the human–robot system's dynamics. This methodology eliminates the need for extensive pre-processing and heuristic delay approximations required in the linear case, streamlining the estimation process while improving precision.

The findings emphasize the NARMAX model's capability to capture intricate nonlinear dynamics and system memory effects, which are essential for modeling human–robot interaction scenarios involving time-varying and adaptive behaviors. The delay values derived using the NARMAX model, consistently within the range of 0.2–0.3 s, align closely with well-established physiological benchmarks for human visual-motor reaction times. This further validates the model's accuracy and applicability in real-world contexts.

Additionally, the frequency-domain analysis introduced in this chapter highlights the model's ability to integrate knowledge of system dynamics, streamlining the estimation process and reducing reliance on pre-processing and heuristic approximations. The NARMAX approach not only replicates but also surpasses the functionality achieved by the linear model, providing a more robust framework for understanding and predicting human control strategies.

These advancements pave the way for future research into adaptive and intelligent robotic systems capable of seamlessly collaborating with humans. By building upon the principles established in this work, future studies can explore more complex scenarios, integrate multimodal sensory inputs, and further enhance the adaptability and responsiveness of human–robot systems. The transition to nonlinear modeling marks a pivotal step in bridging the gap between theoretical models and practical applications, ensuring more effective and intuitive human–robot collaboration.

Chapter 8

Peak-to-Peak Dynamics

In the previous chapters, we introduced the nonlinear autoregressive moving average with exogenous inputs (NARMAX) model as a robust nonlinear approach for modeling human control dynamics in human–robot interaction. While effective, the NARMAX model's complexity can pose challenges in terms of computational cost and scalability, especially in scenarios requiring real-time predictions.

In analyzing the force responses, a recurring pattern of peak values was observed, suggesting the potential to leverage peak-to-peak dynamics (PPD) as a reduced-order model. PPD allows for forecasting system behavior based solely on consecutive peak values, significantly reducing model complexity while retaining predictive accuracy. This chapter explores the application of PPD to the NARMAX model, investigating its feasibility as a simplified yet effective approach for human–robot interaction modeling.

The motivation for this chapter arises from the inherent trade-offs in human–robot interaction systems. As demonstrated in earlier chapters, achieving high accuracy often necessitates increased model complexity, which can hinder the feasibility of real-time deployment. The recurring peaks in force responses during human–robot interaction tasks present a unique opportunity to simplify the model while preserving key predictive capabilities. This chapter aims to investigate PPD as a viable solution for reducing the complexity of the NARMAX model without compromising its performance.

8.1 Peak-to-Peak Dynamics in Complex Systems

A continuous system is said to have PPD if we can determine the value of its next peak starting from the values of the past m peaks, where m is a

127

finite number called "memory." In the presence of peak-to-peak dynamics, the system can be described by reduced-order models, gaining an advantage in computational time.

The PPD property was pointed out for the first time by Lorenz in his famous model of forced dissipative nonperiodic flow (Lorenz, 1963). Lorenz discovered that, given a chaotic system producing an output $y(t)$, there exists a set,

$$S_{\text{PPP}} = (y_k, y_{k-1}), \tag{8.1}$$

called the "peak-to-peak plot" (PPP), which maps the pairs of consecutive peaks in the feature space. Moreover, Lorenz discovered that the blob of points of these peak pairs in the feature space can be approximated with a high degree of accuracy by one or more curves in the relevant plane. This property is strictly related to the geometrical characteristics of the chaotic attractor of the system, which has to be nearly two-dimensional (Strogatz, 1994).

Since Lorenz's first description of this property, systems with PPD have been studied in many fields and formalized in more detail. As explained by Candaten *et al.* (2000), a system is said to present PPD when

$$y_{i+1} = Y(y_i, \ldots, y_{i-m+1}), \tag{8.2}$$

where m is the "memory" number defined above. When the memory factor is 1, PPD is said to be "simple"; in all other cases, it is "complex."

This means that when PPD is present, the occurrence of the next peak of a system, which may be potentially constituted by a considerable number of differential equations, can now be obtained by usually at most two previous peaks, obtaining a reduced-order model.

When PPD is complex, the prediction of the next peak can't be obtained only from the last peak; in this case, extra information is needed. To answer this question, Candaten *et al.* introduced the concept of "pointer," $x_i \in 1, 2, \ldots, k$, which indicates the input states x_i that can influence the y_{i+1} value. If we add this new consideration to the previous PPD definition, we obtain the so-called "peak-to-peak canonical form":

$$y_{i+1} = Y^*(y_i, x_i)$$

$$x_{i+1} = A^*(y_i, x_i), \tag{8.3}$$

Where A^* is a function which is piecewise constant with respect to y_i.

The canonical form equation of PPD tells us that the system describing the dynamics of the peaks is only slightly more complex than a first-order discrete-time system but definitely less complex than a second-order one. On the other hand, it is known that in a generic single-output nth-order system, n samples of the output variable are equivalent to a single sample of the n-dimensional state vector (Takens, 1981). The implication of this is that, as mentioned, the knowledge of a single peak y_i can't be considered sufficient, in general, to forecast the peak y_{i+1}. To do so, we need at least the knowledge of (y_{i-1}, y_i) ($m = 2$ in the first equation). Such consideration is restricted to the dynamics within the system's attractor, i.e., involving the peaks y_{i-1}, y_i, y_{i+1}.

PPD has been successfully applied to state space models with piecewise linearized functional (Candaten and Rinaldi, 2003), Rosenzweig–MacArthur prey–predator model (Candaten and Rinaldi, 2000), and the prey–predator model applied to dynastic cycles (Feichtinger *et al.*, 1996; Piccardi and Feichtinger, 2002).

Piccardi (2008) proposed a parameter estimation method for chaotic systems based on PPD. The method consists of finding the best match between the aprioristic continuous-time model, the 1D map generated with a parameter dependency, and the 1D map derived from the data.

This shows how studying the presence of PPD in a system can be useful for deriving methods and modeling techniques to estimate parameters from data or constructing control methods, even in complex scenarios characterized by chaotic oscillations (Piccardi, 2001).

8.1.1 *Peat-to-peak dynamics in robotics*

The concept of PPD, initially introduced by Lorenz in the context of chaotic systems (Lorenz, 1963), has been extensively explored across various domains. Its utility in capturing essential dynamic behaviors while simplifying complex system dynamics has made it a promising approach for model reduction. Building on Lorenz's findings, Candaten and Rinaldi (2000) formalized PPD, introducing the "memory factor" to quantify the number of past peaks required for accurate prediction of future peaks. This formalization provided a mathematical framework for studying PPD in both deterministic and chaotic systems.

In the domain of robotics, PPD has emerged as a useful technique for addressing challenges in system complexity and computational efficiency.

Robotic systems often involve high-dimensional, nonlinear dynamics that are computationally expensive to model. PPD offers a pathway to reduce model order while maintaining critical performance metrics.

Robotic systems characterized by nonlinear and oscillatory behaviors have seen PPD applied to enhance stability and control. For instance, Zhang *et al.* applied energy-to-peak control for vehicle lateral dynamics, leveraging PPD to stabilize complex systems with reduced computational effort (Zhang *et al.*, 2014). This work demonstrated how PPD could ensure robust performance even under dynamic and uncertain operating conditions.

In the domain of cable-driven robots, Baklouti *et al.* utilized PPD to control vibrations in cable-driven parallel robots (Baklouti *et al.*, 2019). By reducing the peak amplitudes in vibratory responses, they improved the precision of robotic positioning, showcasing the practical advantages of PPD in reducing the impact of oscillatory disturbances.

In soft robotics, Zhao *et al.* applied PPD-inspired techniques to develop sunlight-powered autonomous oscillators for sustainable robotic systems (Zhao *et al.*, 2023a). This innovative application highlighted PPD's potential for optimizing motion dynamics in energy-efficient robotic designs.

One of PPD's most significant contributions to robotics is in model order reduction. By focusing on peak values, PPD enables the simplification of complex dynamics into lower-order representations, making real-time computation more feasible. For example, Nguyen *et al.* demonstrated PPD's utility in simplifying actuator dynamics for vibration reduction in robotic systems, using a sample-and-hold control strategy to ensure stability and reduce oscillatory behavior (Nguyen and Kim, 2023).

Hahn *et al.* investigated the use of PPD for energy optimization in trajectory planning for industrial robots (Hahn and Hosseini, 2024). By smoothing peak-to-peak velocities, they demonstrated significant energy savings while retaining the precision required for complex robotic tasks. This work underscores the broader applicability of PPD in optimizing both performance and resource consumption in robotics.

These examples collectively highlight PPD's potential as a transformative tool for addressing computational challenges in robotics. By applying these principles to the NARMAX model, this chapter aims to extend PPD's utility to human–robot interaction scenarios, focusing on reducing model complexity while maintaining the ability to capture critical dynamic behaviors.

8.2 PPD in Human–Robot Interaction: Methodology

8.2.1 *Adapting PPD to the NARMAX model*

To evaluate the feasibility of PPD in reducing the complexity of the NARMAX model, we restructured the model to predict only the peaks of the force signals. This adaptation leverages the observation that key dynamics of human–robot interaction are captured within these peaks.

The input window for the PPD-adapted model is

$$(y_{i-1}, \ldots, y_{i-n_a}, x_{i-d} + e_{i-1}, \ldots, x_{i-d-n_b} + e_{i-n_c}),$$

where $y_{i-1}, \ldots, y_{i-n_a}$ are outputs from previous peaks and $x_{i-d}, \ldots, x_{i-d-n_b}$ are input states combined with noise elements. The reduced-order model decreased the number of parameters from 10 to 5, significantly improving computational efficiency.

8.2.2 *Dataset preparation and preprocessing*

To train the PPD-adapted model, we modified the original dataset to isolate peak values. This involved the following:

(1) **Rectification and peak extraction**: Force signals were rectified, and local maxima were identified using an iterative algorithm. Thresholds were applied to exclude minor oscillations (minimum amplitude of 10 N and minimum sample distance).

(2) **Data flattening**: The dataset was "flattened" to separate x and y components, resulting in a total of 200 elements for analysis.

(3) **Normalization**: To ensure consistency across experiments, the extracted peaks were normalized before being fed into the model.

To improve readability, the results for all 200 dataset elements are grouped into categories based on performance metrics. Table 8.1 provides a summary with the chosen performance metrics for one in every 10 experiments, while graphical visualizations of predicted versus actual peaks for a subset of experimentally measured (actual) force signals are presented in Figure 8.1.

The performance of the reduced-order model was evaluated using root mean square error (RMSE) and R^2 scores. These metrics quantify the model's predictive accuracy and its ability to replicate the observed force dynamics.

Table 8.1. Results of peak forecasting of the proposed NARMAX model for all the listed elements of the dataset. Root mean square error is expressed in Newton and is calculated by de-normalizing data, while the R2 Score considers normalized data.

Index	RMSE	R2 score
1	0.788339991	0.984102289
10	1.003195685	0.990444132
20	2.201940923	0.972528222
30	1.67177289	0.976332122
40	1.00870708	0.988077678
50	1.58623922	0.985105352
60	2.460776101	0.977637881
70	0.968355499	0.964750538
80	1.528646248	0.966779776
90	2.255182412	0.989172831
100	1.06874344	0.948608385
120	3.841984175	0.979696259
130	1.592419362	0.987348357
140	1.222805204	0.990162584
150	1.764032442	0.987473683
160	3.98407289	0.985796984
170	1.383867045	0.977346606
180	1.723981168	0.964440686
190	2.413074624	0.991536574
200	0.663964704	0.961805728

8.3 Results and Discussion

8.3.1 *Predictive accuracy of the reduced-order model*

The results, summarized in Table 8.1, demonstrate that the PPD-adapted model achieved high predictive accuracy with minimal degradation compared to the full-order NARMAX model. RMSE values averaged 1.34 N, while R^2 scores exceeded 0.98 across the dataset.

The results confirm the feasibility of using PPD to simplify the NARMAX model for human–robot interaction scenarios. Despite a reduction in model order, the predictive accuracy remained high, with RMSE values averaging 1.34 N and R^2 scores exceeding 0.98.

The reduced-order model offers several advantages:

- **Computational efficiency**: The reduction in model order halved the computational time required for training and inference.

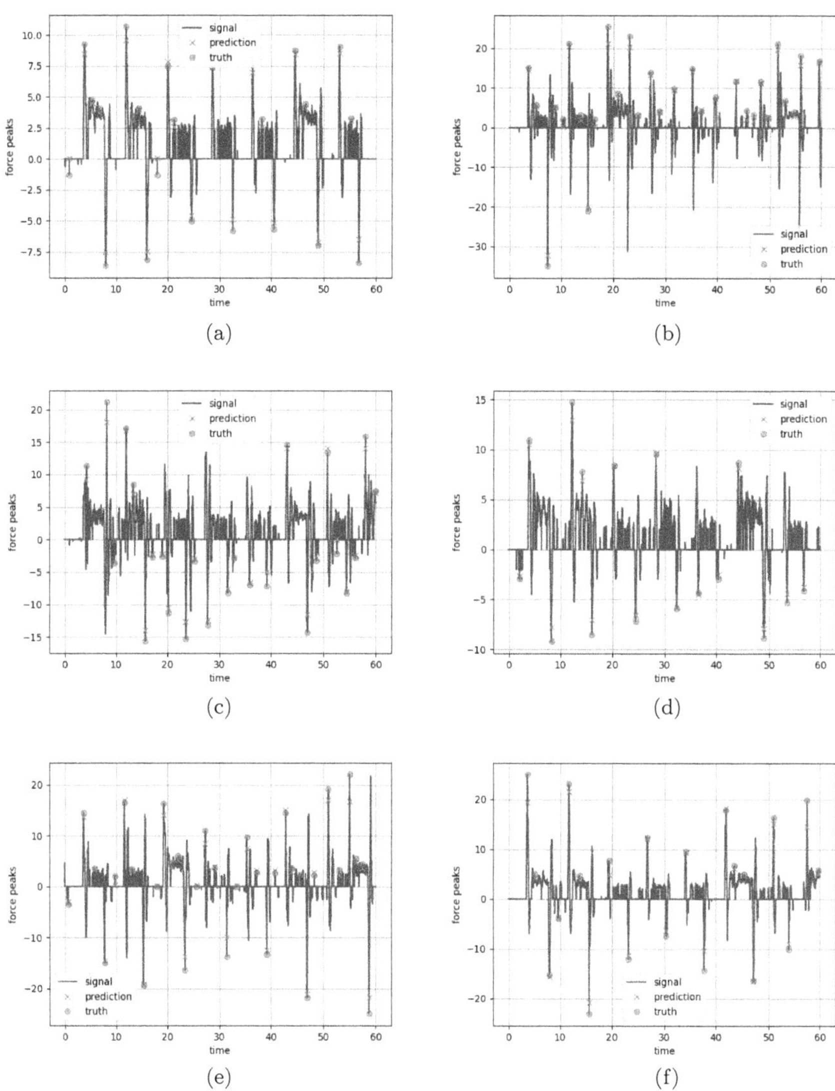

Fig. 8.1. Measured vs. predicted peak values (respectively indicated as dots and crosses) displayed in a summary of 18 experiments extracted (1 every 10) from the original dataset of 200 listed elements (100 for x and 100 for y signal components).

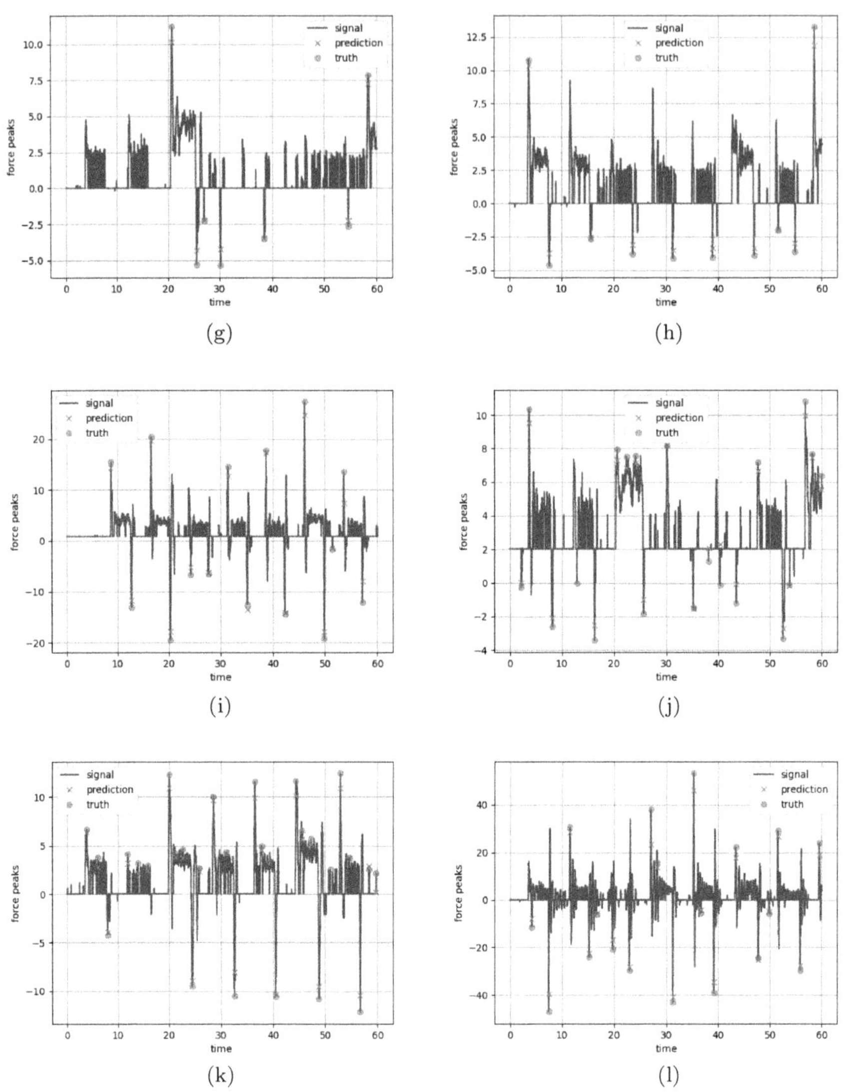

Fig. 8.1. (*Continued*)

- **Interpretability**: By focusing on peak values, the model provides a clearer representation of system dynamics, facilitating parameter estimation and control design.
- **Scalability**: The simplified structure enhances the model's applicability to real-time scenarios.

Figure 8.1 illustrates predicted versus ground truth peak values for a representative subset of the dataset. Predicted values are marked with crosses, while ground truth values are depicted with circles. The high overlap between predicted and actual peaks demonstrates the accuracy of the reduced-order model.

8.4 Final Considerations

This chapter demonstrates the potential of PPD as a reduced-order modeling technique for human–robot interaction. By applying PPD to the NARMAX model, we achieved a significant reduction in complexity while maintaining high predictive accuracy. These findings pave the way for future research into real-time modeling and control applications, leveraging the efficiency and simplicity of PPD.

Conclusion

In this book, a vast variety of human models aiming to represent their control behavior when interacting with a controlled machine were investigated and, most importantly, used to create custom modeling approaches inspired by such vast state-of-the-art techniques that were applied in the context of human–robot interaction.

In the first part of this work, the linear modeling techniques were investigated. Such approaches proved to be able to represent well human physiological control districts, which are involved in the perception, control, and actuation processes. Moreover, we saw how such physiological representations can be linked to higher-level control-theory-fashioned models like the crossover model, structural model, and optimal control model.

The aforementioned linear approaches, with particular reference to the crossover model, were used to propose a *precision model of the human–robot complex* when performing a cooperative manipulation task with a collaborative robot, which, in the first proposed experimental study, was Universal Robot's UR5 manipulator. The proposed model proved to be able to accurately represent human reaction when an external forcing function is applied to the system. This consideration was verified from two perspectives: the experimental identification of human reaction delay time with particular attention to intra-subject constancy and inter-subject variability and the model's capability to simulate well the position vector of the human–robot complex offline after the experiment.

In the work's second part, we focused on nonlinear dynamics and modeling efforts directed toward their representation. Nonlinear models proved to be able to represent complex processes involved in the human control strategy, such as decision-making, the creation of a long-term strategy able to overcome short-term goals, and the ability of humans to transform their behavior from a linear one to a nonlinear impulsive one when the complexity level of the demanded task increases.

Such techniques were studied in the first place in a different application scenario: transport systems. Such a field, in fact, is one of the first in which artificial intelligence frameworks were successfully applied to represent human decision-making and its ability to deal with complex problems. This led to the development of systems able to perform automatic lane-changing, as mentioned, finding a long-term strategy based on a reinforcement learning framework, and the control of quadrotors facing highly nonlinear external dynamics.

In the study of nonlinear state-of-the-art models, a focal point of our study was how data-driven and model-based techniques can be combined to represent partially unknown dynamics. Such consideration led us to the definition of a nonlinear autoregressive moving average with exogenous inputs (NARMAX) model, able to describe human response with a precision level that would not be possible with any linear technique.

The proposed NARMAX model was able to simulate and forecast human control behavior and estimate its intention during an interaction task with a compliant collaborative robot. The model was trained and tested in the first place with the same experimental data acquired and used in the previously proposed linear model. This time, the model was used to represent human force output, which, from a control block scheme point of view, is the robot's input. Highly impulsive components characterize such a force signal; it is noisy and present high variability between each experiment and each subject.

Despite this, both the RMSE and the R2 score value testify to the outstanding model performance when forecasting x and y force components at each instant. The plots also testify to this consideration, where the ground truth and the prediction signals are superimposed almost entirely.

To verify model generalization capabilities, it was tested with new data produced by a different experimental setup and using a different robot, the Comau NS16. This time, a further step was performed: To let the robot know the human intention nearly in real time, the model was used to forecast human-generated force online. Position and force signals were acquired in order to let the model forecast the x and y components of the force reference that the human will generate after the next 12 ms, which is the time required by the model to compute a new prediction.

With respect to state-of-the-art approaches like those adopted by Cremer *et al.* (2019) and Suzuki and Furuta (2012), the proposed modeling and control techniques proved to be able, despite the simplicity of their structures, to replicate and forecast highly nonlinear and impulsive output dynamics of the human control behavior during a continuous interaction.

This consideration is confirmed by the generated force plots, where the predicted force samples reproduce the ground truth vector with high fidelity at each instant. The "real-time" performance of the proposed framework proved to have a slow initial tuning phase and a reliable and fast second phase, where the time delay became almost negligible.

The robot high-level controller showed some difficulties in distinguishing human-generated reference force, which had to be converted to a position reference to let the robot anticipate human motion from unwanted external contacts. The state-machine approach alone has some limitations in performing such a complex duty, which will have to be performed by another dedicated framework. Despite this, in the few cases in which the human reference was recognized, the advantages seemed to be important by looking at both force outputs and the position tracking error.

Aside from the human model, the robotic platform's model can also be approximated. In this case, the admittance controller and its virtual mass–spring–damper system can be modeled as a second-order transfer function. The knowledge of both the system's main components allowed us to extract the information about control delay from empirical data, obtaining a mean delay value of 0.3 s. The optimization and identification techniques were relatively simple when compared to more complex approaches which have been proposed (Boer and Kenyon, 1998; Karami *et al.*, 2021) when less information on the system dynamics was available.

The optimal model used to obtain predictions at every time frame in which the human interacted with the robot was a NARMAX model of the 10th order. However, as known, several interesting techniques have been proposed in the past to lower model order and complexity. In particular, peak-to-peak dynamics (PPD) have been successfully studied and applied in the past for complex systems characterized by chaotic behavior. To exploit their applicability in our model, the latter was retrained to forecast only the peak values of the system's output, but reducing its order as much as possible without excessively affecting its accuracy. The results showed prediction performance comparable to the ones obtained with the complete-order model, but the order was reduced from 10 to 5, further reducing model complexity and required computational time. To the best of our knowledge, this was the first attempt to study PPD in NARMAX models.

Appendix

A.1 ARMAX Models

The autoregressive moving average with exogenous inputs (ARMAX) model is a generalization of time-series modeling techniques that includes both autoregressive and moving average components while also incorporating external inputs. ARMAX models are widely used in system identification, econometrics, and engineering to model, predict, and control dynamic systems.

The ARMAX model is particularly useful when the system's output is influenced not only by its past values and random disturbances but also by external input variables. This model can be represented as a discrete-time difference equation or in the transfer function form.

The ARMAX model is represented mathematically as

$$y(t) + \sum_{i=1}^{n_a} a_i y(t-i) = \sum_{j=1}^{n_b} b_j u(t-j) + \sum_{k=1}^{n_c} c_k e(t-k) + e(t), \qquad \text{(A.1)}$$

where:

- $y(t)$ is the output variable at time t,
- $u(t)$ is the input variable (exogenous input) at time t,
- $e(t)$ is a white noise error term (innovation),
- a_i are the coefficients of the autoregressive (AR) terms,
- b_j are the coefficients of the exogenous input (X) terms,
- c_k are the coefficients of the moving average (MA) terms,
- $n_a, n_b,$ and n_c are the orders of the AR, X, and MA components, respectively.

The ARMAX model can also be expressed in terms of transfer functions:

$$A(q)y(t) = B(q)u(t) + C(q)e(t), \tag{A.2}$$

where:

- $A(q) = 1 + a_1 q^{-1} + a_2 q^{-2} + \cdots + a_{n_a} q^{-n_a}$,
- $B(q) = b_1 q^{-1} + b_2 q^{-2} + \cdots + b_{n_b} q^{-n_b}$,
- $C(q) = 1 + c_1 q^{-1} + c_2 q^{-2} + \cdots + c_{n_c} q^{-n_c}$,
- q^{-1} is the backward shift operator, such that $q^{-1}y(t) = y(t-1)$.

Equation (A.2) illustrates the input–output relationship in terms of polynomials in the backward shift operator q^{-1}.

The parameters a_i, b_j, and c_k are typically estimated using methods such as the following:

- **Least squares (LS)**: minimizes the sum of squared prediction errors.
- **Maximum likelihood estimation (MLE)**: Maximizes the likelihood of observing the data.
- **Recursive methods**: suitable for online or adaptive modeling.

The parameter estimation process often involves solving a nonlinear optimization problem, as the ARMAX model is linear in parameters but nonlinear in error terms.

The main components constituting the model are the autoregressive part, the moving average, and the exogenous input.

The autoregressive component models the dependency of the current output $y(t)$ on its past values:

$$A(q)y(t) = y(t) + \sum_{i=1}^{n_a} a_i y(t-i). \tag{A.3}$$

The moving average component models the effect of past errors $e(t)$ on the current output:

$$C(q)e(t) = e(t) + \sum_{k=1}^{n_c} c_k e(t-k). \tag{A.4}$$

The exogenous input component captures the effect of external inputs $u(t)$ on the output:

$$B(q)u(t) = \sum_{j=1}^{n_b} b_j u(t-j). \tag{A.5}$$

A.2 Artificial Neural Networks

Artificial neural networks (ANNs) are computational models inspired by the biological neural networks that constitute the human brain. They are designed to recognize patterns, model complex relationships, and perform tasks such as classification, regression, clustering, and decision-making. ANNs are fundamental building blocks in machine learning and are widely used in applications ranging from computer vision to natural language processing.

A.2.1 *Basic structure of ANNs*

An ANN consists of interconnected layers of nodes (neurons). The three primary types of layers are as follows:

- **Input layer**: receives input data and passes it to the subsequent layers.
- **Hidden layers**: intermediate layers that perform computations to extract features and learn patterns.
- **Output layer**: produces the final output, such as a classification label or a predicted value.

Each neuron in a layer is connected to neurons in the adjacent layer, and these connections are associated with weights (w_{ij}) that determine the influence of one neuron on another. A bias term (b_j) is also included to adjust the output.

A.2.2 *Mathematical representation*

The operation of a neuron can be expressed mathematically as

$$z_j = \sum_i w_{ij} x_i + b_j,$$

where:

- x_i is input from the previous layer.
- w_{ij} is the weight connecting input i to neuron j.
- b_j is the bias term.
- z_j is a linear combination of inputs, weights, and bias.

The neuron applies an activation function $f(z_j)$ to introduce nonlinearity:

$$a_j = f(z_j).$$

Common activation functions include the following:

- **Sigmoid**: $f(z) = \frac{1}{1+e^{-z}}$, which maps outputs to the range $(0, 1)$.
- **ReLU (Rectified linear unit)**: $f(z) = \max(0, z)$, which introduces sparsity.
- **Tanh**: $f(z) = \frac{e^z - e^{-z}}{e^z + e^{-z}}$, mapping outputs to $(-1, 1)$.
- **Softmax**: used in classification tasks to convert raw scores into probabilities.

A.2.3 *Forward propagation*

During forward propagation, input data are passed through the network layer by layer. At each layer, the weighted sum of inputs is calculated, and the activation function is applied. This process continues until the output layer produces the final result. Mathematically,

$$\mathbf{a}^{(l+1)} = f(\mathbf{W}^{(l)}\mathbf{a}^{(l)} + \mathbf{b}^{(l)}),$$

where:

- $\mathbf{a}^{(l)}$: activations from layer l.
- $\mathbf{W}^{(l)}$: weight matrix for layer l.
- $\mathbf{b}^{(l)}$: bias vector for layer l.

A.2.4 *Error calculation*

The error or loss quantifies the difference between the network's output and the ground truth. Common loss functions include the following:

- **Mean squared error (MSE)**: used for regression tasks:

$$\text{MSE} = \frac{1}{n} \sum_{i=1}^{n} (\hat{y}_i - y_i)^2.$$

- **Cross-entropy loss**: used for classification tasks:

$$\text{Cross-Entropy} = -\sum_{i=1}^{n} y_i \log(\hat{y}_i).$$

A.2.5 *Training an ANN: Backpropagation and optimization*

Training an ANN involves adjusting the weights and biases to minimize the loss function. This is achieved through an iterative process using backpropagation and an optimization algorithm.

A.2.5.1 *Backpropagation*

Backpropagation computes the gradient of the loss function with respect to each weight and bias in the network. It involves two steps:

- **Forward pass**: Calculate the network's output and the loss.
- **Backward pass**: Propagate the error backward through the network to compute gradients. Using the chain rule, the gradient of the loss with respect to a weight w_{ij} is

$$\frac{\partial \text{Loss}}{\partial w_{ij}} = \frac{\partial \text{Loss}}{\partial z_j} \cdot \frac{\partial z_j}{\partial w_{ij}}.$$

A.2.5.2 *Stochastic gradient descent (SGD)*

SGD is an optimization algorithm that updates weights and biases using the computed gradients:

$$w_{ij} \leftarrow w_{ij} - \eta \frac{\partial \text{Loss}}{\partial w_{ij}},$$

where η is the learning rate.

Variants of SGD include the following:

- **Momentum**: accelerates convergence by considering the past gradient:

$$v_t = \beta v_{t-1} - \eta \nabla \text{Loss}, \quad w_{ij} \leftarrow w_{ij} + v_t.$$

- **Adam**: combines momentum and adaptive learning rates:

$$m_t = \beta_1 m_{t-1} + (1 - \beta_1) \nabla \text{Loss}, \quad v_t = \beta_2 v_{t-1} + (1 - \beta_2)(\nabla \text{Loss})^2$$

$$\hat{m}_t = \frac{m_t}{1 - \beta_1^t}, \quad \hat{v}_t = \frac{v_t}{1 - \beta_2^t},$$

$$w_{ij} \leftarrow w_{ij} - \eta \frac{\hat{m}_t}{\sqrt{\hat{v}_t} + \epsilon}.$$

A.2.6 *Convergence and overfitting*

Effective training requires careful tuning of hyperparameters such as the learning rate, batch size, and number of epochs. Overfitting, where the network memorizes training data instead of generalizing, can be mitigated using the following:

- **Regularization**: Add penalties to the loss function, such as L_1 or L_2 regularization.

- **Dropout**: Randomly deactivate neurons during training to prevent reliance on specific pathways.
- **Early stopping**: Halt training when the validation error stops improving.

A.3 Robot Kinematics

Robot kinematics deals with the motion of a robot without considering the forces that cause this motion. It is divided into *forward kinematics* and *inverse kinematics*.

A.3.1 *Forward kinematics*

Forward kinematics calculates the position and orientation of the robot's end-effector based on joint parameters. For a serial manipulator with n degrees of freedom, the position $\mathbf{P}$ and orientation $\mathbf{R}$ of the end-effector are expressed as

$$\mathbf{T} = \prod_{i=1}^{n} \mathbf{A}_i, \tag{A.6}$$

where $\mathbf{A}_i$ represents the transformation matrix of the ith joint. A typical $\mathbf{A}_i$ matrix is given by the Denavit–Hartenberg (DH) parameters as

$$\mathbf{A}_i = \begin{bmatrix} \cos\theta_i & -\sin\theta_i\cos\alpha_i & \sin\theta_i\sin\alpha_i & a_i\cos\theta_i \\ \sin\theta_i & \cos\theta_i\cos\alpha_i & -\cos\theta_i\sin\alpha_i & a_i\sin\theta_i \\ 0 & \sin\alpha_i & \cos\alpha_i & d_i \\ 0 & 0 & 0 & 1 \end{bmatrix}. \tag{A.7}$$

A.3.2 *Inverse kinematics*

Conversely, inverse kinematics (IK) determines joint parameters from the desired position and orientation of the end-effector. This involves solving the following problem:

$$\mathbf{T}_{\text{desired}} = \prod_{i=1}^{n} \mathbf{A}_i. \tag{A.8}$$

Solving this equation can be challenging due to multiple solutions or singularities; for this reason, it is often done using numerical solving algorithms.

In other words, solving inverse robot kinematics involves finding joint configurations for a robot that achieve a desired end-effector pose (position and orientation) in Cartesian space. This problem is generally nonlinear and can have multiple solutions, no solutions, or even infinite solutions depending on the robot's kinematics and the desired pose. Several numerical methods are commonly employed to solve this problem, including the following.

A.3.3 *Jacobian-based methods*

These iterative methods rely on the robot's Jacobian matrix, which maps joint velocities onto end-effector velocities. They are widely used due to their generality and ability to handle redundant and under-actuated robots.

A.3.3.1 *Newton–Raphson method*

This method uses the Jacobian to linearize the problem locally and then Updates the joint configuration iteratively:

$$\Delta \mathbf{q} = \mathbf{J}^{\dagger}(\mathbf{q})(\mathbf{x}_d - \mathbf{x}(\mathbf{q})),$$

where:

- $\Delta \mathbf{q}$ is the joint update,
- $\mathbf{J}^{\dagger}$ is the pseudo-inverse of the Jacobian,
- $\mathbf{x}_d$ is the desired pose,
- $\mathbf{x}(\mathbf{q})$ is the current pose.

This method requires the Jacobian to be non-singular near the solution.

A.3.3.2 *Damped least squares (DLS)*

This method modifies the pseudo-inverse to handle singularities by adding a damping term:

$$\mathbf{J}^{\dagger} = \mathbf{J}^T (\mathbf{J}\mathbf{J}^T + \lambda^2 \mathbf{I})^{-1},$$

where λ is the damping factor. DLS improves numerical stability and avoids large joint updates near singularities.

A.3.4 *Optimization-based methods*

Optimization techniques treat IK as a constrained optimization problem where the objective is to minimize an error function, such as the difference between the desired and current end-effector pose.

A.3.4.1 *Gradient descent*

This method iteratively updates joint variables in the direction of the negative gradient of a cost function $f(\mathbf{q})$:

$$\mathbf{q}_{k+1} = \mathbf{q}_k - \alpha \nabla f(\mathbf{q}_k),$$

where α is the learning rate.

A.3.4.2 *Nonlinear programming (NLP)*

NLP solves IK as an optimization problem,

$$\text{Minimize } f(\mathbf{q}) = \|\mathbf{x}_d - \mathbf{x}(\mathbf{q})\|^2,$$

subject to constraints such as joint limits:

$$\mathbf{q}_{\min} \leq \mathbf{q} \leq \mathbf{q}_{\max}.$$

Algorithms like sequential quadratic programming (SQP) and interiorpoint methods are commonly used.

A.3.5 *Cyclic coordinate descent (CCD)*

CCD is a heuristic iterative method that adjusts one joint at a time to reduce the error between the current and desired end-effector positions. For each joint i:

- compute the vector from the joint to the end-effector and the desired position;
- rotate the joint to align these vectors as closely as possible.

This method is simple to implement and computationally efficient for low-dimensional problems, but it may converge slowly for complex robots.

A.3.6 *Numerical root-finding methods*

In these methods, we solve the kinematic equations as a system of nonlinear equations:

$$\mathbf{x}_d - \mathbf{x}(\mathbf{q}) = 0$$

Common algorithms include the following:

- **Newton–Raphson**: uses the Jacobian to iteratively refine the solution.
- **Broyden's method**: A quasi-Newton method that approximates the Jacobian, reducing computational cost.

A.3.7 *Heuristic and metaheuristic methods*

These methods are useful for highly redundant robots or complex kinematic structures. Examples include the following:

- **Genetic algorithms (GAs)**: uses evolutionary principles to explore the solution space.
- **Particle swarm optimization (PSO)**: models the solution process as a swarm of particles searching for the best configuration.
- **Simulated annealing (SA)**: Mimics the cooling process in metallurgy to find global minima.

A.3.8 *Hybrid methods*

Hybrid methods combine multiple approaches to exploit their respective strengths. For example, we can use optimization methods to get close to a solution and refine it with Jacobian-based techniques.

A.3.9 *Closed-form solutions (analytical methods)*

Although not a numerical method, closed-form solutions are important when they exist. For specific robot configurations (e.g., 6-DOF manipulators with spherical wrists), analytical methods can provide explicit solutions. Numerical methods are typically used when closed-form solutions are unavailable or impractical.

A.3.10 *Practical considerations*

- **Redundancy handling**: For redundant robots, prioritize solutions by adding constraints or secondary objectives, such as minimizing energy or avoiding obstacles.
- **Singularities**: Special care must be taken near singular configurations, where the Jacobian loses rank.
- **Real-time feasibility**: Numerical methods must be efficient enough for real-time applications in control and teleoperation.

Each method has trade-offs in terms of speed, accuracy, and robustness, and the choice of method depends on the robot's kinematic structure, task requirements, and computational resources.

A.4 Robot Dynamics

Robot dynamics involves the forces and torques that cause motion. The equations of motion can be derived using Newton–Euler or Lagrangian methods.

One commonly used method is the Newton–Euler algorithm. This approach calculates forces and torques by considering linear and angular accelerations. The dynamic equation for a link i is given by

$$\mathbf{F}_i = m_i \mathbf{a}_i, \quad \boldsymbol{\tau}_i = \mathbf{I}_i \boldsymbol{\alpha}_i + \boldsymbol{\omega}_i \times (\mathbf{I}_i \boldsymbol{\omega}_i), \tag{A.9}$$

where $\mathbf{F}_i$ and $\boldsymbol{\tau}_i$ are the force and torque, respectively, $\mathbf{a}_i$ is the linear acceleration, $\boldsymbol{\alpha}_i$ is the angular acceleration, m_i is the mass, and $\mathbf{I}_i$ is the inertia tensor.

A different approach is provided by using the Lagrangian method. In particular, the Lagrangian is defined as

$$\mathcal{L} = K - U, \tag{A.10}$$

where K is the kinetic energy and U is the potential energy. The equations of motion are derived from

$$\frac{d}{dt}\left(\frac{\partial \mathcal{L}}{\partial \dot{q}_i}\right) - \frac{\partial \mathcal{L}}{\partial q_i} = \tau_i, \tag{A.11}$$

where q_i and $\dot{q}_i$ are the joint position and velocity respectively, and τ_i is the joint torque.

A.5 Impedance Control

Impedance control is a widely used control strategy in robotics to regulate the dynamic interaction between a robot and its environment. Instead of directly controlling position or force, impedance control governs the relationship between these variables, allowing the robot to behave as a virtual mechanical system characterized by desired mass, damping, and stiffness.

Impedance control is based on the concept of mimicking a desired mechanical impedance:

$$\mathbf{F}_{\text{ext}} = \mathbf{M}_d\ddot{\mathbf{x}} + \mathbf{B}_d\dot{\mathbf{x}} + \mathbf{K}_d(\mathbf{x}_d - \mathbf{x}), \tag{A.12}$$

where:

- $\mathbf{F}_{\text{ext}}$ is the external force exerted by the environment;
- $\mathbf{M}_d$, $\mathbf{B}_d$, and $\mathbf{K}_d$ are the desired mass, damping, and stiffness matrices, respectively;
- $\mathbf{x}_d$ and $\mathbf{x}$ are the desired and actual positions of the robot in Cartesian space, respectively;
- $\dot{\mathbf{x}}$ and $\ddot{\mathbf{x}}$ are the velocity and acceleration of the robot in Cartesian space, respectively.

The goal of impedance control is to design the control input such that the robot behaves according to the desired dynamics described in equation (A.12).

A.5.1 *Joint space impedance control*

In joint space impedance control, the desired impedance is defined in terms of joint positions, velocities, and accelerations. The control law is given by

$$\tau = \tau_d + \mathbf{M}_d(\ddot{\mathbf{q}}_d - \ddot{\mathbf{q}}) + \mathbf{B}_d(\dot{\mathbf{q}}_d - \dot{\mathbf{q}}) + \mathbf{K}_d(\mathbf{q}_d - \mathbf{q}), \tag{A.13}$$

where:

- τ is the control torque applied to the joints;
- τ_d is the feedforward torque to compensate for gravity and other dynamic effects;
- $\mathbf{q}_d$ and $\mathbf{q}$ are the desired and actual joint positions;
- $\dot{\mathbf{q}}_d$ and $\dot{\mathbf{q}}$ are the desired and actual joint velocities;
- $\ddot{\mathbf{q}}_d$ and $\ddot{\mathbf{q}}$ are the desired and actual joint accelerations.

This formulation ensures that the joint behavior matches the desired impedance dynamics defined by the matrices $\mathbf{M}_d$, $\mathbf{B}_d$, and $\mathbf{K}_d$.

A.5.2 *Cartesian space impedance control*

In Cartesian space impedance control, the control objective is to regulate the impedance at the robot's end-effector. The control law can be expressed as

$$\mathbf{F}_{\mathrm{cmd}} = \mathbf{M}_d(\ddot{\mathbf{x}}_d - \ddot{\mathbf{x}}) + \mathbf{B}_d(\dot{\mathbf{x}}_d - \dot{\mathbf{x}}) + \mathbf{K}_d(\mathbf{x}_d - \mathbf{x}), \tag{A.14}$$

where:

- $\mathbf{F}_{\mathrm{cmd}}$ is the commanded force at the end-effector;
- $\mathbf{x}_d$, $\dot{\mathbf{x}}_d$, and $\ddot{\mathbf{x}}_d$ are the desired position, velocity, and acceleration in Cartesian space, respectively;
- $\mathbf{x}$, $\dot{\mathbf{x}}$, and $\ddot{\mathbf{x}}$ are the actual position, velocity, and acceleration in Cartesian space, respectively.

The desired impedance matrices $\mathbf{M}_d$, $\mathbf{B}_d$, and $\mathbf{K}_d$ define how the end-effector interacts with the environment.

To apply Cartesian impedance control on a robot, the relationship between joint torques and end-effector forces must be established using the robot's Jacobian matrix $\mathbf{J}(\mathbf{q})$:

$$\tau = \mathbf{J}^T(\mathbf{q})\mathbf{F}_{\mathrm{cmd}}. \tag{A.15}$$

A.5.3 *Joint vs. Cartesian impedance control*

Joint impedance control is computationally simpler and is often used for tasks that require precise control of joint motion. However, it lacks direct control over end-effector behavior in Cartesian space, making it less intuitive for interaction tasks.

Cartesian impedance control is more suitable for tasks involving direct interaction with the environment, such as force-guided assembly or human–robot collaboration. It provides intuitive control over end-effector dynamics but requires accurate modeling of the robot's kinematics and dynamics.

A.5.4 *Applications in robotics*

Impedance control is widely used in various robotic applications:

- **Industrial robots**: for tasks involving contact, such as polishing or grinding.
- **Humanoid robots**: to ensure safe and compliant interaction with humans.
- **Rehabilitation robotics**: for assisting patients during physical therapy by adapting to their movements.
- **Teleoperation**: to enhance the operator's control by providing haptic feedback.

A.5.5 *Implementation challenges*

Despite its advantages, implementing impedance control faces several challenges:

- **Parameter tuning**: The choice of $\mathbf{M}_d$, $\mathbf{B}_d$, and $\mathbf{K}_d$ significantly affects system performance and stability.
- **Dynamic modeling**: Accurate models of the robot's dynamics are required for effective control.
- **Sensor noise**: Measurement errors in force and position sensors can degrade performance.

A.6 Robot Operating System

The robot operating system (ROS) is a flexible and modular framework designed to facilitate the development of robotic applications. It is not an actual operating system but a middleware that provides essential tools and libraries for building and managing robot software. ROS enables developers to focus on high-level functionality without worrying about low-level hardware details, making it a widely adopted standard in academia and industry.

A.6.1 *Overview of ROS architecture*

ROS follows a distributed architecture that allows various components of a robotic system to communicate seamlessly. The main components of ROS are as follows:

- **Nodes**: Fundamental building blocks of a ROS system, nodes are processes that perform computations. Each node is designed to execute a specific task, such as sensing, control, or decision-making.

- **Topics**: These are used for message-based communication between nodes. Topics follow a publish-subscribe model where publishers send data to a topic and subscribers receive it.
- **Services**: These enable request-response communication between nodes. Services are used when immediate feedback is required, such as triggering an action.
- **Actions**: Actions are similar to services but designed for long-running tasks. They allow feedback and preemption during execution.
- **Parameter server**: A parameter server is a centralized storage for configuration parameters accessible by all nodes.
- **ROS master**: This component manages the registration of nodes and facilitates communication between them.

A.6.2   *Key features of ROS*

ROS encourages modular design by allowing developers to create reusable packages. A package is a collection of nodes, configuration files, and other resources that work together to perform a specific function.

Moreover, it provides a hardware abstraction layer that enables the same software to work across different hardware platforms. This is achieved through device drivers and standard communication interfaces.

Its communication system is based on inter-process communication mechanisms:

$$\text{Communication Type} \in \{\text{Topics, Services, Actions}\}.$$

ROS uses message types defined by users or provided by standard libraries to exchange data efficiently.

A.6.2.1   *Simulation and visualization tools*

- **Gazebo**: s powerful 3D simulation tool integrated with ROS for testing algorithms and robot designs in realistic environments.
- **RViz**: s visualization tool used for debugging and monitoring robot states, sensor data, and trajectories.

A.6.2.2   *Extensive libraries*

The framework includes a rich set of libraries for motion planning, perception, control, and more. Popular libraries include the following:

- **MoveIt**: for motion planning and manipulation.

- **navigation stack**: for autonomous navigation.
- **tf**: for managing coordinate transformations.

ROS primarily supports programming in Python and C++, with additional support for languages such as Java and MATLAB. Each ROS node is implemented as a separate program, and communication is established through ROS-specific APIs.

Bibliography

Abbas, M. Z., Sajjad, I. A., Hussain, B., Liaqat, R., Rasool, A., Padmanaban, S., and Khan, B. (2022). An adaptive-neuro fuzzy inference system based-hybrid technique for performing load disaggregation for residential customers, *Scientific Reports* **12**(1), 1–14.

Abbas, W., Abouelatta, O. B., El-Azab, M., Elsaidy, M., Megahed, A. A., *et al.* (2010). Optimization of biodynamic seated human models using genetic algorithms, *Engineering* **2**(09), 710–720.

Adigbli, P., Grand, C., Mouret, J.-B., and Doncieux, S. (2007). Nonlinear attitude and position control of a micro quadrotor using sliding mode and backstepping techniques, in *7th European Micro Air Vehicle Conference and Flight Competition (MAV07)*, pp. 1–9.

Afloare, A. I. and Ionita, A. (2014). Prediction of the handling qualities and pilot-induced oscillation rating levels, *INCAS Bulletin* **6**, 3–13.

Aghabayk, K., Forouzideh, N., and Young, W. (2013). Exploring a local linear model tree approach to car-following, *Computer-Aided Civil and Infrastructure Engineering* **28**(8), 581–593.

Aguirre, L. A., Rodrigues, G. G., and Jácome, C. R. (1998). Identificaçao de sistemas nao lineares utilizando modelos narmax polinomiais–uma revisao e novos resultados, *SBA Controle e Automação* **9**(2), 90–106.

Ajoudani, A., Tsagarakis, N., and Bicchi, A. (2012). Tele-impedance: Teleoperation with impedance regulation using a body–machine interface, *The International Journal of Robotics Research* **31**(13), 1642–1656.

Akanyeti, O., Nehmzow, U., and Iglesias, R. (2008). Robot training using system identification, *Robotics and Autonomous Systems* **57**, 619–630.

Akash, D., Piyush, K. K., and Nitin, D. C. (2021). A study to target energy consumption in wastewater treatment plant using machine learning algorithms, in *31st European Symposium on Computer Aided Process Engineering*, pp. 1511–1516.

Alabdullah, M. H. and Abido, M. A. (2022). Microgrid energy management using deep q-network reinforcement learning, *Alexandria Engineering Journal* **61**(11), 9069–9078.

AlMomani, A. A. R., Sun, J., and Bollt, E. (2020). How entropic regression beats the outliers problem in nonlinear system identification, *Chaos: An Interdisciplinary Journal of Nonlinear Science* **30**, 1.

Andreassi, J. (2010). *Psychophysiology: Human Behavior and Physiological Response* (Taylor and Francis, New York).

Anya, A. R., Rouphail, N. M., Frey, H. C., and Schroeder, B. (2014). Application of aimsun microsimulation model to estimate emissions on signalized arterial corridors, *Transportation Research Record* **2428**(1), 75–86.

Arruda, J., Yagi, Y., Coburn, K., and Estes, K. (1999). Effects of aerobic exercise and gender on visual and auditory p300, reaction time, and accuracy, *European Journal of Applied Physiology and Occupational Physiology* **80**(5), 402–408.

Aung, Y. M., Anam, K., and Al-Jumaily, A. (2015). Continuous prediction of shoulder joint angle in real-time, in *2015 7th International IEEE/EMBS Conference on Neural Engineering (NER)* (IEEE), pp. 755–758.

Avci, E. and Akpolat, Z. H. (2006). Speech recognition using a wavelet packet adaptive network based fuzzy inference system, *Expert Systems with Applications* **31**(3), 495–503.

Munyazikwiye, B. B., Vysochinskiy, D., Khadyko, M., and G. Robbersmyr, K. (2018). Prediction of vehicle crashworthiness parameters using piecewise lumped parameters and finite element models, *Designs* **2**(4), 43–59.

Bachelder, E. N. and Aponso, B. (2022). Human pilot control adaptation: A physiological interpretation, *AIAA SCITECH 2022 Forum*, pp. 2446–2458. DOI: 10.2514/6.2022-2446.

Baklouti, S., Courteille, E., Lemoine, P., and Caro, S. (2019). Vibration reduction of cable-driven parallel robots through elasto-dynamic model-based control, *Mechanism and Machine Theory* **137**, 354–370.

Baron, S., Kleinman, D., and Levison, W. (1970). An optimal control model of human response part II: Prediction of human performance in a complex task, *Automatica* **6**(3), 371–383.

Baron, S. and Kleinman, D. L. (1969). The human as an optimal controller and information processor, *IEEE Transactions on Man-Machine Systems* **10**(1), 9–17.

Bartolomaeus, W., Boutaib, Y., Nestler, S., and Rauhut, H. (2022). Path classification by stochastic linear recurrent neural networks, *Advances in Continuous and Discrete Models*, 13.

Bentin, S. and McCarthy, G. (1994). The effects of immediate stimulus repetition on reaction time and event-related potentials in tasks of different complexity, *Journal of Experimental Psychology: Human Perception and Performance* **20**(1), 85–94.

Bergman, J. (2016). Connectome: How the brain's wiring makes us who we are, *Journal of Interdisciplinary Studies* **28**(1–2), 192–194.

Berthoz, A., Pavard, B., and Young, L. R. (1975). Perception of linear horizontal self-motion induced by peripheral vision (line-arvection) basic characteristics and visual-vestibular interactions, *Experimental Brain Research* **23**(5), 471–489.

Bertsekas, D. (2012). *Dynamic Programming and Optimal Control: Volume I*, Vol. 1 (Athena Scientific, Belmont).

Billings, S. and Coca, D. (2008). Robot training using system identification, *Robotics and Autonomous Systems* **56**(6), 766–779.

Billings, S. A. (2013). *Nonlinear System Identification: NARMAX Methods in the Time, Frequency, and Spatio-Temporal Domains* (John Wiley & Sons, Chichester).

Bizzi, E., Accornero, N., Chapple, W., and Hogan, N. (1984). Posture control and trajectory formation during arm movement, *Journal of Neuroscience* **4**, 2738–2744.

Boer, E. R. and Kenyon, R. V. (1998). Estimation of time-varying delay time in nonstationary linear systems: an approach to monitor human operator adaptation in manual tracking tasks, *IEEE Transactions on Systems, Man, and Cybernetics-Part A: Systems and Humans* **28**(1), 89–99.

Bouabdallah, S., Noth, A., and Siegwart, R. (2004). Pid vs lq control techniques applied to an indoor micro quadrotor, in *2004 IEEE/RSJ International Conference on Intelligent Robots and Systems (IROS)*, Vol. 3, pp. 2451–2456.

Bouabdallah, S. and Siegwart, R. (2005). Backstepping and sliding-mode techniques applied to an indoor micro quadrotor, in *Proceedings of the 2005 IEEE International Conference on Robotics and Automation*, pp. 2247–2252.

Bryson, A. E. and Ho, Y.-C. (2018). *Applied Optimal Control: Optimization, Estimation, and Control* (Routledge, New York).

Bucolo, M., Buscarino, A., Fortuna, L., and Gagliano, S. (2020). Bifurcation scenarios for pilot induced oscillations, *Aerospace Science and Technology* **106**, 106194–106199.

Burdet, E., Tee, K. P., Mareels, I., Milner, T. E., Chew, C. M., Franklin, D. W., and Kawato, M. (2006). Stability and motor adaptation in human arm movements, *Biological Cybernetics* **94**(1), 20–32.

Cai, C., Schroeder, M., and Liu, S. (2024). A mpc-based approach for motion planning on redundant manipulators in human robot collaboration, in *IEEE International Conference on Automation Science and Engineering*, pp. 3387–3392.

Caianiello, M., Ricci, M., Smaldone, A., Hussain, S., Iacono, C., and Ficuciello, F. (2024). Optimizing safety and efficiency in the suturing task: A comparison of model predictive control and control barrier function framework, in *Proceedings of IEEE Workshop on Advanced Robotics and Its Social Impacts, ARSO*, pp. 104–109.

Candaten, M. and Rinaldi, S. (2000). Peak-to-peak dynamics: A critical survey, *International Journal of Bifurcation and Chaos* **10**, 08, 1805–1819.

Candaten, M. and Rinaldi, S. (2003). Peak-to-peak dynamics in food chain models, *Theoretical Population Biology* **63**(4), 257–267.

Castillo, P., Albertos, P., Garcia, P., and Lozano, R. (2006). Simple real-time attitude stabilization of a quad-rotor aircraft with bounded signals, in

Proceedings of the 45th IEEE Conference on Decision and Control, pp. 1533–1538.

Chen, F., Xie, W., and Xia, T. (2020). Target tracking algorithm based on kernel correlation filter with anti-occlusion mechanisms, in *15th IEEE International Conference on Signal Processing (ICSP)*, Vol. 1 (IEEE), pp. 220–225.

Chen, R., Jin, X., Laima, S., Huang, Y., and Li, H. (2022). Intelligent modeling of nonlinear dynamical systems by machine learning, *International Journal of Non-Linear Mechanics* **142**, 103984–103995.

Chen, S. and Billings, S. A. (1989). Representations of non-linear systems: The narmax model, *International Journal of Control* **49**(3), 1013–1032.

Cheng, L., Wang, Z., Jiang, F., and Li, J. (2021). Adaptive neural network control of nonlinear systems with unknown dynamics, *Advances in Space Research* **67**(3), 1114–1123.

Coe, T. E., Xing, J. T., Shenoi, R. A., and Taunton, D. (2009). A simplified 3-d human body-seat interaction model and its applications to the vibration isolation design of high-speed marine craft, *Ocean Engineering* **36**(9), 732–746.

Cowling, I. D., Yakimenko, O. A., Whidborne, J. F., and Cooke, A. K. (2010). Direct method based control system for an autonomous quadrotor, *Journal of Intelligent & Robotic Systems* **60**(2), 285–316.

Cremer, S., Das, S. K., Wijayasinghe, I. B., Popa, D. O., and Lewis, F. L. (2019). Model-free online neuroadaptive controller with intent estimation for physical human–robot interaction, *IEEE Transactions on Robotics* **36**(1), 240–253.

Csáji, B. C. *et al.* (2001). Approximation with artificial neural networks, *Faculty of Sciences, Etvs Lornd University, Hungary* **24**(48), 7.

Curry, R., Young, L., Hoffman, W., and Kugel, D. (1976). A pilot model with visual and motion cues, in *Proceedings of the AIAA Visual and Motion Simulation Conference*.

da Silva, Y. M., Andrade, F. A., Sousa, L., de Castro, G. G., Dias, J. T., Berger, G., Lima, J., and Pinto, M. F. (2022). Computer vision based path following for autonomous unammed aerial systems in unburied pipeline onshore inspection, *Drones* **6**(12), 410–429.

Dai, Y., Chen, Q., Zhang, J., Wang, X., Chen, Y., Gao, T., Xu, P., Chen, S., Liao, S., Jiang, H., *et al.* (2022). Enhanced oblique decision tree enabled policy extraction for deep reinforcement learning in power system emergency control, *Electric Power Systems Research* **209**, 107932–107943.

D'Anniballe, V. M., Tushar, F. I., Faryna, K., Han, S., Mazurowski, M. A., Rubin, G. D., and Lo, J. Y. (2022). Multi-label annotation of text reports from computed tomography of the chest, abdomen, and pelvis using deep learning, *BMC Medical Informatics and Decision Making* **22**(1), pp. 1–12.

Das, A., Lewis, F., and Subbarao, K. (2009). Backstepping approach for controlling a quadrotor using lagrange form dynamics, *Journal of Intelligent and Robotic Systems* **56**(1), 127–151.

De Groote, F. and Falisse, A. (2022). Rapid predictive simulations to study the interaction between motor control and musculoskeletal dynamics in healthy and pathological human movement, in *Converging Clinical and Engineering Research on Neurorehabilitation IV: Proceedings of the 5th International Conference on Neurorehabilitation (ICNR2020), October 13–16, 2020* (Springer), pp. 327–331.

Demir, U., Kocaoğlu, S., and Akdoğan, E. (2016). Human impedance parameter estimation using artificial neural network for modelling physiotherapist motion, *Biocybernetics and Biomedical Engineering* **36**(2), 318–326.

Desideri, L., Ottaviani, C., Malavasi, M., di Marzio, R., and Bonifacci, P. (2019). Emotional processes in human-robot interaction during brief cognitive testing, *Computers in Human Behavior* **90**, 331–342.

Desplenter, T. and Trejos, A. L. (2018). Evaluating muscle activation models for elbow motion estimation, *Sensors* **18**(4), 1004.

DeVore, R., Hanin, B., and Petrova, G. (2021). Neural network approximation, *Acta Numerica* **30**, 327–444.

Dey, S., Reang, N. M., Majumder, A., Deb, M., and Das, P. K. (2020). A hybrid ann-fuzzy approach for optimization of engine operating parameters of a ci engine fueled with diesel-palm biodiesel-ethanol blend, *Energy* **202**, 117813–117830.

Dieterich, O., Götz, J., Vu, D., Haverdings, B., Masarati, P., and Pavel, M. (2008). Adverse rotorcraft-pilot coupling: Recent research activities in Europe.

Dikmen, İ. C., Arisoy, A., and Temeltas, H. (2009). Attitude control of a quadrotor, in *Proceedings of the 4th International Conference on Recent Advances in Space Technologies*, pp. 722–727.

Doman, D. (1999). Optimal control pilot modeling for resolving cooper-harper rating discrepancies, in *AIAA Atmospheric Flight Mechanics Conference and Exhibit* (AIAA-1999-4091, Portland, OR), pp. 176–186.

Doman, D. B. and Anderson, M. R. (2000). A fixed–order optimal control model of human operator response, *Automatica* **36**(3), pp. 409–418.

Dovom, E. M., Azmoodeh, A., Dehghantanha, A., Newton, D. E., Parizi, R. M., and Karimipour, H. (2019). Fuzzy pattern tree for edge malware detection and categorization in IoT, *Journal of Systems Architecture* **97**, 1–7.

Du, Y., Wang, Y., and Chan, C.-Y. (2014). Autonomous lane-change controller via mixed logical dynamical, in *Proceedings of the 17th International IEEE Conference on Intelligent Transportation Systems (ITSC)*, pp. 1154–1159.

Dubois, D. M. (2002). Theory of computing anticipatory systems based on differential delayed-advanced difference equations, in *AIP Conference Proceedings*, Vol. 627 (American Institute of Physics), pp. 3–16.

Dydek, Z. T., Annaswamy, A. M., and Lavretsky, E. (2012). Adaptive control of quadrotor uavs: A design trade study with flight evaluations, *IEEE Transactions on Control Systems Technology* **21**(4), 1400–1406.

Edkins, C. R. (1994). The prediction of pilot opinion ratings using optimal and suboptimal pilot models, Vol. AD-278 679 (Wright-Patterson Air Force Base, OH).

Efremov, A. V., Ogloblin, A. V., and Koshelenko, A. V. (1998). Evaluation and prediction of aircraft handling qualities, in *23rd Atmospheric Flight Mechanics Conference*, AIAA-98-4145.

Efremov, A. V. and Tjaglik, M. S. (2011). The development of perspective displays for highly precise tracking tasks, in *Advances in Aerospace Guidance, Navigation and Control*, pp. 163–174.

Elfring, J., Van De Molengraft, R., and Steinbuch, M. (2014). Learning intentions for improved human motion prediction, *Robotics and Autonomous Systems* **62**(4), 591–602.

Erlhagen, W., Mukovskiy, A., Bicho, E., Panin, G., Kiss, C., Knoll, A., Van Schie, H., and Bekkering, H. (2006). Goal-directed imitation for robots: A bio-inspired approach to action understanding and skill learning, *Robotics and Autonomous Systems* **54**(5), 353–360.

Ertugrul, S. (2008). Predictive modeling of human operators using parametric and neuro-fuzzy models by means of computer-based identification experiment, *Engineering Applications of Artificial Intelligence* **21**(2), 259–268.

Falisse, A., Pitto, L., Kainz, H., Hoang, H., Wesseling, M., Van Rossom, S., Papageorgiou, E., BarOn, L., Hallemans, A., Desloovere, K., *et al.* (2020). Physics–based simulations to predict the differential effects of motor control and musculoskeletal deficits on gait dysfunction in cerebral palsy: A retrospective case study, *Frontiers in Human Neuroscience* **14**, 40.

Fayazi, S. A., Wan, N., Lucich, S., Vahidi, A., and Mocko, G. (2013). Optimal pacing in a cycling time-trial considering cyclist's fatigue dynamics, in *American Control Conference*, pp. 6442–6447.

Feichtinger, G., Forst, C. V., and Piccardi, C. (1996). A nonlinear dynamical model for the dynastic cycle, *Chaos, Solitons & Fractals* **7**(2), 257–271.

Feldman, A. G. (1966). Functional tuning of nervous system with control of movement or maintenance of a steady posture. II. Controllable parameters of the muscles, *Biophysics* **11**, 565–578.

Feldman, A. G., Goussev, V., Sangole, A., and Levin, M. F. (2007). Threshold position control and the principle of minimal interaction in motor actions, *Progress in Brain Research* **165**, 267–281.

Feldman, A. G. and Levin, M. F. (2009). The equilibrium–point hypothesis-past, present and future, *Progress in Motor Control: A Multidisciplinary Perspective*, pp. 699–726.

Fernandez, C. and Goldberg, J. M. (1971). Physiology of peripheral neurons innervating semicircular canals of the squirrel monkey. II. Response to sinusoidal stimulation and dynamics of peripheral vestibular system, *Journal of Neurophysiology* **34**(4), 661–675.

Fischer, B. and Weber, H. (1997). Effects of stimulus conditions on the performance of antisaccades in man, *Experimental Brain Research* **116**(2), 191–204.

Flash, T. (1987). The control of hand equilibrium trajectories in multi-joint arm movements, *Biological Cybernetics* **57**, 257–274.

Fruggiero, F., Lambiase, A., Panagou, S., and Sabattini, L. (2020). Cognitive human modeling in collaborative robotics, *Procedia Manufacturing* **51**, 584–591.

Gai, E. G. and Curry, R. E. (1976). A model of the human observer in failure detection tasks, *IEEE Transactions on Systems, Man, and Cybernetics, SMC* **6**(2), 85–94.

Gao, M. and Sun, T. (2020). Predictive control for assistive robotic systems based on human force dynamics, *Journal of Rehabilitation Robotics* **12**, 202.

Gao, Q., Li, J., Zhu, Y., Wang, S., Liufu, J., and Liu, J. (2023). Hand gesture teleoperation for dexterous manipulators in space station by using monocular hand motion capture, *Acta Astronautica* **204**, 630–639.

Gawthrop, P., Loram, I., Lakie, M., and Gollee, H. (2011). Intermittent control: A computational theory of human control, *Biological Cybernetics* **104**(1), 31–51.

George, G. R. (2008). *New Methods of Mathematical Modeling of Human Behavior in the Manual Tracking Task* (State University of New York at Binghamton).

Gestwa, M. and Grigorie, L. (2011). Using fuzzy control for modeling the control behaviour of a human pilot, *Fuzzy Controllers, Theory and Applications*, pp. 297–326 (Intech, Rijeka).

Gibson, J. (1997). The Definition, Understanding and Design of Aircraft Handling Qualities, Delft University Press.

Gomi, H. and Kawato, M. (1996). Equilibrium-point control hypothesis examined by measured arm-stiffness during multi-joint movement, *Science* **272**, 117–120.

Grant, O. L., Stol, K. A., Swain, A., and Hess, R. A. (2015). Handling qualities of a twin ducted-fan aircraft: An analytical evaluation, *Journal of Guidance, Control, and Dynamics* **38**(6), 1126–1131.

Grant, P. R., Yam, B., Hosman, R., and Schroeder, J. A. (2005). The effect of simulator motion on pilot's control behavior for helicopter yaw control tasks, in *AIAA Modeling and Simulation Technologies Conference* (AIAA).

Gribble, P. L. and Ostry, D. J. (1999). Compensation for interaction torques during single and multi-joint limb movement, *Journal of Neurophysiology* **82**, 2310–2326.

Griffin, M. J. (2001). The validation of biodynamic models, *Clinical Biomechanics* **16**(1), S81–S92.

Groen, E. L., Smaili, M. H., and Hosman, R. J. (2007). Perception model analysis of flight simulator motion for a decrab maneuver, *Journal of Aircraft* **44**(2), 427–435.

Gupta, A. and Smith, J. (2022). Data-driven techniques for predictive control in robotics: A review, *Robotics and Autonomous Systems* **142**, 104357.

Hahn, I. and Hosseini, S. (2024). Analyzing the estimation of energy consumption in the trajectory planning of an industrial robot, *IEEE Transactions on Industrial Electronics* **71**, 1453–1465.

Haji Hassani, R., Bannwart, M., Bolliger, M., Seel, T., Brunner, R., and Rauter, G. (2022). Real-time motion onset recognition for robot-assisted gait rehabilitation, *Journal of NeuroEngineering and Rehabilitation* **19**(1), 1–14.

Halligan, P. W., Zeman, A., and Berger, A. (1999). *Phantoms in the Brain.* https://www.bmj.com/content/319/7210/587.

Han, J., Ding, Q., Xiong, A., and Zhao, X. (2015a). A state-space emg model for the estimation of continuous joint movements, *IEEE Transactions on Industrial Electronics* **62**(7), 4267–4275.

Han, J., Ding, Q., Xiong, A., and Zhao, X. (2015b). A state-space emg model for the estimation of continuous joint movements, *IEEE Transactions on Industrial Electronics* **62**(7), 4267–4275.

Han, S., Zhu, H., Zhong, Q., Shi, K., and Kwon, O.-M. (2023). Adaptive event-triggered fuzzy positioning control for unmanned marine vehicles with actuator saturation and hybrid attacks, *IEEE Transactions on Fuzzy Systems*, 1–13.

Hashemi, A. and Vikalo, H. (2018). Accelerated orthogonal least-squares for large-scale sparse reconstruction, *Digital Signal Processing* **82**, 91–105.

He, B., Yuan, H., Meng, J., and Gao, S. (2020). Brain–computer interfaces, *Neural Engineering*, 131–183.

Heerspink, H., Berkouwer, W., Stroosma, O., van Paassen, R., Mulder, M., and Mulder, B. (2005). Evaluation of vestibular thresholds for motion detection in the simona research simulator, in *AIAA Modeling and Simulation Technologies Conference and Exhibit*, p. 6502.

Heffley, R. (2010). Use of a task-pilot-vehicle (tpv) model as a tool for flight simulator math model development, in *AIAA Guidance Navigation, and Control Conference* (AIAA, Toronto, ON, Canada).

Henriques, J., Caseiro, R., Martins, P., and Batista, J. (2015). High-speed tracking with kernelized correlation filters, *IEEE Transactions on Pattern Analysis and Machine Intelligence* **37**, 583–596.

Henson, M. A. and Seborg, D. E. (1997). Feedback linearizing control, *Nonlinear Process Control* **4**, 149–231.

Hess, R. (2018). Human–in–the–loop control, in *Control System Applications* (CRC Press, Boca Raton), pp. 327–334.

Hess, R. A. (1972). Optimal control approximations for time delay systems, *AIAA Journal* **10**(11), 1536–1538.

Hess, R. A. (1977). Prediction of pilot opinion ratings using an optimal pilot model, *Human Factors: The Journal of the Human Factors and Ergonomics Society* **19**(5), 459–475.

Hess, R. A. (1979). Structural model of the adaptive human pilot, *Journal of Guidance, Control, and Dynamics* **3**(5), 416–423.

Hess, R. A. (1980a). A pilot modeling technique for handling-qualities research, in *6th Atmospheric Flight Mechanics Conference*.

Hess, R. A. (1980b). Structural model of the adaptive human pilot, *Journal of Guidance and Control* **3**(5), 416–423.

Hess, R. A. (1984). Effects of time delays on systems subject to manual control, *Journal of Guidance, Control, and Dynamics* **7**(4), 416–421.

Hess, R. A. (1989). Theory for aircraft handling qualities based upon a structural pilot model, *Journal of Guidance, Control, and Dynamics* **12**(6), 792–797.

Hess, R. A. (1990a). Analyzing manipulator and feel system effects in aircraft flight control, *IEEE Transactions on Systems, Man, and Cybernetics* **20**(4), 923–931.

Hess, R. A. (1990b). Model for human use of motion cues in vehicular control, *Journal of Guidance, Control, and Dynamics* **13**(3), 476–482.

Hess, R. A. (1997). Unified theory for aircraft handling qualities and adverse aircraft–pilot coupling, *Journal of Guidance, Control, and Dynamics* **20**(6), 1141–1148.

Hess, R. A. (2005). Rudder control strategies and force/feel system designs in transport aircraft, *Journal of Guidance, Control, and Dynamics* **28**(6), 1251–1262.

Hess, R. A. (2008). Obtaining multi-loop pursuit-control pilot models from computer simulation, in *Proceedings of the Institution of Mechanical Engineers (Journal of Aerospace Engineering, Part G)*, 189–199.

Hess, R. A. (2010). Multi–axis pilot modelling: Models and methods for wake vortex encounter simulations, in *Presentation to WakeNet–Europe Safety Workshop* **3**.

Hess, R. A. and Joyce, R. (2013). Analytical investigation of transport aircraft handling qualities, in *AIAA Atmospheric Flight Mechanics Conference* 19–22.

Hess, R. A. and Malsbury, T. (1991). Closed-loop assessment of flight simulator fidelity, *Journal of Guidance, Control, and Dynamics* **14**(1), 191–197.

Hess, R. A., Malsbury, T., and Atencio Jr., A. (1993). Flight simulator fidelity assessment in a rotorcraft lateral translation maneuver, *Journal of Guidance, Control, and Dynamics* **16**(1), 79–85.

Hess, R. A. and Siwakosit, W. (2001). Assessment of flight simulator fidelity in multiaxis tasks including visual cue quality, *Journal of Aircraft* **38**(4), 607–614.

Hess, R. A. and Sunyoto, I. (1985). Toward a unifying theory for aircraft handling qualities, *Journal of Guidance, Control, and Dynamics* **8**(4), 40–46.

Hess, R. A. (1979). A rationale for human operator pulsive control behavior, *Journal of Guidance and Control* **2**(3), 221–227.

Hiatt, L. M., Narber, C., Bekele, E., Khemlani, S. S., and Trafton, J. G. (2017). Human modeling for human-robot collaboration, *The International Journal of Robotics Research* **36**(5–7), 580–596.

Higgins, I. (2021). Generalizing universal function approximators, *Nature Machine Intelligence* **3**(3), 192–193.

Hoehne, G. (1999). A biomechanical pilot model for prediction of roll ratcheting, in *AIAA Atmospheric Flight Mechanics Conference* (AIAA), pp. 187–196.

Hinton, G. E. and Salakhutdinov, R. R. (2006). Reducing the dimensionality of data with neural networks, *Science* **313**(5786), 504–507.

Hirakawa, A., Asano, J., Sato, H., and Teramukai, S. (2018). Master protocol trials in oncology: Review and new trial designs, *Contemporary Clinical Trials Communications* **12**, 1–8.

Hochreiter, S. and Schmidhuber, J. (1997). Long short-term memory, *Neural Computation* **9**(8), 1735–1780.

Hoffmann, G., Waslander, S., and Tomlin, C. (2008). Quadrotor helicopter trajectory tracking control, in *AIAA Guidance, Navigation and Control Conference and Exhibit*, pp. 7410–7424.

Hornik, K. (1991). Approximation capabilities of multilayer feedforward networks, *Neural Networks* **4**(2), 251–257.

Hosman, R. (1996). Pilot's Perception and Control of Aircraft Motions (Delft University Press).

Hosman, R., Schuring, J., and Van der Geest, P. (2005). Pilot model development for the manual balked landing maneuver, in *AIAA Modeling and Simulation Technologies Conference and Exhibit*, AIAA-2005-5884.

Hosman, R. and Stassen, H. (1999). Pilot's perception in the control of aircraft motions, *Control Engineering Practice* **7**(11), 1421–1428.

Hosman, R. and Van der Vaart, J. C. (1981). Effects of vestibular and visual motion perception on task performance, *Acta Psychologica* **48**(1), 271–287.

Huang, T. and Fu, R. (2022). Prediction of the driver's focus of attention based on feature visualization of a deep autonomous driving model, *Knowledge-Based Systems* **251**, 109006–109016.

Huber, M., Rickert, M., Knoll, A., Brandt, T., and Glasauer, S. (2008). Human-robot interaction in handing-over tasks, in *RO-MAN 2008-The 17th IEEE International Symposium on Robot and Human Interactive Communication* (IEEE), pp. 107–112.

Hussain, N., Ali, S., Ovinis, M., and Arshad, M. (2019). Underactuated coupled nonlinear adaptive control synthesis using u-model for multivariable unmanned marine robotics, *IEEE Transactions on Control Systems Technology* **27**, 319–334.

Hwang, C.-L. and Abebe, H. B. (2021). Generalized and heterogeneous nonlinear dynamic multiagent systems using online rnn-based finite-time formation tracking control and application to transportation systems, *IEEE Transactions on Intelligent Transportation Systems* **23**(8), 13708–13720.

Innocenti, M. (1988). The Optimal Control Pilot Model and Applications (Advisory Group for Aerospace Research and Development Neuilly-Sur-Seine, France).

Iranmanesh, H., Keshavarz, M., and Abdollahzade, M. (2016). Predicting dust storm occurrences with local linear neuro fuzzy model: A case study in Ahvaz city, Iran, in *International Conference on Soft Computing-MENDEL*, pp. 158–167.

Ito, M. (2000). Mechanisms of motor learning in the cerebellum, *Brain Research* **886**(1–2), 237–245.

Iwamoto, M., Kisanuki, Y., Watanabe, I., Furusu, K., Miki, K., and Hasegawa, J. (2002). Development of a finite element model of the total human model for safety (thums) and application to injury reconstruction, in *Proceedings of the International IRCOBI Conference*, pp. 18–20.

Jagacinski, R., Flach, J., and Erlbaum, L. (2004). *Control Theory for Humans: Quantitative Approaches to Modeling Performance* (CRC Press, Boca Raton).

Jang, J.-S. (1993). Anfis: Adaptive-network-based fuzzy inference system, *IEEE Transactions on Systems, Man, and Cybernetics* **23**(3), 665–685.

Jiang, H.-T., Su, X.-D., and Li, H. (2014). Dynamic multi-attribute decision making based on advantage retention degree, *International Journal of Control and Automation* **7**(9), 389–398.

Jiang, L., Chen, D., Li, Z., and Wang, Y. (2022). Risk representation, perception, and propensity in an integrated human lane-change decision model, *IEEE Transactions on Intelligent Transportation Systems* **23**(12), 23474–23487.

Jin, I. G. and Orosz, G. (2018). Connected cruise control among human-driven vehicles: Experiment–based parameter estimation and optimal control design, *Transportation Research Part C: Emerging Technologies* **95**, 445–459.

Johannsen, G. and Rouse, W. B. (1979). Mathematical concepts for modeling human behavior in complex man-machine systems, *Human Factors* **21**(6), 733–747.

Johansen, T. A. and Foss, B. (1993). Constructing narmax models using armax models, *International Journal of Control* **58**(5), 1125–1153.

Johnston, D. E. and Aponso, B. L. (1988). Design considerations of manipulator and feel system characteristics in roll tracking, NASA CR-4111 .

Joos, A., Hoffmann, M., and Stein, T. (2018). Human center of mass trajectory models using nonlinear model predictive control, *IFAC-PapersOnLine* **51**, 366–371.

Jordan, M. I. and Rumelhart, D. E. (2013). Forward models: Supervised learning with a distal teacher, in *Backpropagation* (Psychology Press), pp. 189–236. https://doi.org/10.1207/s15516709cog1603_1.

Junior, W., Martins, S. A. M., and Nepomuceno, E. G. (2021). Meta-model structure selection: Building polynomial narx model for regression and classification, *arXiv preprint* arXiv:2109.09917.

Kam, H. R., Lee, S.-H., Park, T., and Kim, C.-H. (2015). Rviz: A toolkit for real domain data visualization, *Telecommunication Systems* **60**, 337–345.

Karami, M., Abbasi, M., and Vossoughi, G. (2021). Design of a continuous fractional-order nonsingular terminal sliding mode control with time delay estimation for fes control of human knee joint, in *2021 7th International Conference on Control, Instrumentation and Automation (ICCIA)* (IEEE), pp. 1–5.

Karamouzas, I. and Overmars, M. (2010). A velocity-based approach for simulating human collision avoidance, in *Intelligent Virtual Agents: 10th International Conference, IVA 2010, Philadelphia, PA, USA, September 20–22, 2010. Proceedings 10* (Springer), pp. 180–186.

Kashiri, N., Baccelliere, L., Muratore, L., Laurenzi, A., Ren, Z., Hoffman, E. M., Kamedula, M., Rigano, G. F., Malzahn, J., Cordasco, S., *et al.* (2019). Centauro: A hybrid locomotion and high power resilient manipulation platform, *IEEE Robotics and Automation Letters* **4**(2), 1595–1602.

Katayama, M., Inoue, S., and Kawato, M. (1998). A strategy of motor learning using adjustable parameters for arm movement, in *Proceedings of the 20th Annual International Conference of the IEEE Engineering in Medicine and Biology Society* (IEEE, Hong Kong), pp. 2370–2373.

Katayama, M. and Kawato, M. (1993). Virtual trajectory and stiffness ellipse during multijoint arm movement predicted by neural inverse models, *Biological Cybernetics* **69**, 353–362.

Kawato, M. (1999). Internal models for motor control and trajectory planning, *Current Opinion in Neurobiology* **9**(6), 718–727.

Kawato, M., Furukawa, K., and Suzuki, R. (1987). A hierarchical neural–network model for control and learning of voluntary movement, *Biological Cybernetics* **57**, 169–185.

Khoramshahi, M. and Billard, A. (2019). A dynamical system approach to task–adaptation in physical human–robot interaction, *Autonomous Robots* **43**, 927–946.

Kitazaki, S. and Griffin, M. J. (1997a). A modal analysis of whole-body vertical vibration, using a finite element model of the human body, *Journal of Sound and Vibration* **200**(1), 83–103.

Kitazaki, S. and Griffin, M. J. (1997b). Resonance behaviour of the seated human body and effects of posture, *Journal of Biome-Chanics* **31**(2), 143–149.

Kleinman, D. (1969). Optimal lineer control for systems with time delay and observation noise, *IEE Transactions on Automatic Control* **14**(4).

Kleinman, D. L., Baron, S., and Levison, W. (1970). An optimal control model of human response part i: Theory and validation, *Automatica* **6**(3), 357–369.

Kleinman, D. L. and Killingsworth, W. R. (1974). A predictive pilot model for stol aircraft landing, NASA CR–2374 .

Konnaris, M. A., Brendel, M., Fontana, M. A., Otero, M., Ivashkiv, L. B., Wang, F., and Bell, R. D. (2022). Computational pathology for musculoskeletal conditions using machine learning: Advances, trends, and challenges, *Arthritis Research & Therapy* **24**(1), 1–15.

Kordnoori, S., Sharifi, A., and Shah-Hosseini, H. (2022). Human fall detection using neuro-fuzzy models based on ensemble learning, *Progress in Artificial Intelligence* **11**, 1–14.

Kormushev, P., Calinon, S., and Caldwell, D. G. (2010). Reinforcement learning in physical human-robot interaction, in *IEEE International Conference on Robotics and Automation (ICRA)*, pp. 362–367.

Kraskov, A., Stögbauer, H., and Grassberger, P. (2004). Estimating mutual information, *Physical Review E* **69**(6), 066138.

Kumar, A. and Ghosh, I. (2022). Non-compliance behavior of pedestrians and the associated conflicts at signalized intersections in india, *Safety Science* **147**(10560), 4.

Kyriacou, T., Akanyeti, O., Nehmzow, U., and Iglesias, R. (2006). Visual task identification using polynomial models, *Proceedings of Autonomous Robotic Systems*.

Lacerda, W. R., da Andrade, L. P. C., Oliveira, S. C. P., and Martins, S. A. M. (2020). Sysidentpy: A python package for system identification using narmax models, *Journal of Open Source Software* **5**(54), p. 2384.

Langner, R., Steinborn, M., Chatterjee, A., and Sturm, W. (2010). Mental fatigue and temporal preparation in simple reaction-time performance, *Acta Psychologica* **133**(1), 64–75.

Lasota, P. A., Fong, T., Shah, J. A., *et al.* (2017). A survey of methods for safe human-robot interaction, *Foundations and Trends® in Robotics* **5**(4), 261–349.

Latash, M. L. and Gottlieb, G. L. (1991). Reconstruction of shifting elbow joint compliant characteristics during fast and slow movements, *Neuroscience* **43**, 697–712.

Lawrence, D. A. (1993). Stability and transparency in bilateral teleoperation, *IEEE Transactions on Robotics and Automation* **9**(5), 624–637.

LeCun, Y., Bengio, Y., and Hinton, G. (2015). Deep learning, *Nature* **521**(7553), 436–444.

Lee, D., Jin Kim, H., and Sastry, S. (2009). Feedback linearization vs. adaptive sliding mode control for a quadrotor helicopter, *International Journal of Control, Automation and Systems* **7**(3), 419–428.

Lee, H. and Kim, H. J. (2017). Trajectory tracking control of multirotors from modelling to experiments: A survey, *International Journal of Control, Automation and Systems* **15**(1), 281–292.

Levine, W. S. (2018). *The Control Handbook* (three volume Set) (CRC Press, Boca Raton).

Levison, W. H. (1989). Alternative treatments of attention-sharing within the optimal control model, in *Systems, Man and Cybernetics Conference Proceedings* (IEEE), pp. 744–749.

Levison, W. H. and Harrah, C. B. (1977). Biomechanical and performance response of man in six different directional axis vibration environments, Technical report AMRL-TR-77-71, Aerospace Medical Research Laboratory.

Li, C., Zheng, P., Li, S., Pang, Y., and Lee, C. K. (2022). Ar-assisted digital twin-enabled robot collaborative manufacturing system with human-in-the-loop, *Robotics and Computer-Integrated Manufacturing* **76**, 102321–102330.

Li, J., Wang, R., and Pan, L. (2023). An enhanced emg-driven musculoskeletal model based on non-negative matrix factorization, *Biomedical Signal Processing and Control* **79**.

Li, K., Zhang, J., Liu, X., and Zhang, M. (2019a). Estimation of continuous elbow joint movement based on human physiological structure, *Biomedical Engineering Online* **18**, pp. 1–15.

Li, K., Zhang, J., Zhang, H., and Zhang, M. (2018). A novel method for estimating continuous motion and time-varying stiffness of human elbow joint, in *2018 IEEE International Conference on Robotics and Biomimetics (ROBIO)* (IEEE), pp. 299–304.

Li, T., Wu, J., and Chan, C.-Y. (2019b). Evolutionary learning in decision making for tactical lane changing, in *2019 IEEE Intelligent Transportation Systems Conference (ITSC)*, pp. 1826–1831.

Li, W. and Todorov, E. (2004). Iterative linear quadratic regulator design for nonlinear biological movement systems, in *Proceedings of the 1st International Conference on Informatics in Control, Automation and Robotics*, pp. 222–229.

Li, W. and Todorov, E. (2006). An iterative optimal control and estimation design for nonlinear stochastic system, in *Proceedings of the 45th IEEE Conference on Decision and Control*, pp. 3242–3247.

Li, X.-D., Ho, J. K., and Chow, T. W. (2005). Approximation of dynamical time-variant systems by continuous-time recurrent neural networks, *IEEE Transactions on Circuits and Systems II: Express Briefs* **52**(10), 656–660.

Li, W. and Todorov, E. (2007). Iterative linearization methods for approximately optimal control and estimation of non-linear stochastic system, *International Journal of Control* **80**(9), pp. 1439–1453.

Liu, J. Z., Brown, R. W., and Yue, G. H. (2002). A dynamical model of muscle activation, fatigue, and recovery, *Biophysical Journal* **82**(5), 2344–2359.

Liu, W., Kim, S.-W., Pendleton, S., and Ang, M. H. (2015). Situation-aware decision making for autonomous driving on urban road using online pomdp, *2015 IEEE Intelligent Vehicles Symposium (IV)*, pp. 1126–1133.

Lone, M. and Cooke, A. (2014). Review of pilot models used in aircraft flight dynamics, *Aerospace Science and Technology* **34**, 55–74.

Lone, M. M. (2013). Pilot modeling for airframe loads analysis, Ph.D. thesis, University of Cranfield.

Lone, M. M. and Cooke, A. K. (2012). Pilot-model-in-the-loop simulation environment to study large aircraft dynamics, *Proceedings of the Institution of Mechanical Engineers, Part G: Journal of Aerospace Engineering* **227**(3), 555–568.

Lorenz, E. N. (1963). Deterministic nonperiodic flow, *Journal of Atmospheric Sciences* **20**(2), 130–141.

Luo, W., Xing, J., Milan, A., Zhang, X., Liu, W., and Kim, T. K. (2021). Multiple object tracking: A literature review, *Artificial Intelligence* **293**(10344), 8.

Ma, L., Chablat, D., Bennis, F., and Zhang, W. (2009). A new simple dynamic muscle fatigue model and its validation, *International Journal of Industrial Ergonomics* **39**(1), 211–220.

Ma, W., Xue, H., Sun, X., Mao, S., Wang, L., Liu, Y., Wang, Y., and Lin, X. (2022). A novel multi-branch hybrid neural network for motor imagery eeg signal classification, *Biomedical Signal Processing and Control* **77**, 103718–103735.

Madani, T. and Benallegue, A. (2006). Control of a quadrotor mini-helicopter via full state backstepping technique, in *Proceedings of the 45th IEEE Conference on Decision and Control*, pp. 1515–1520.

Magdaleno, R. E. and Mc Ruer, D. T. (1971). Experimental validation and analytical elaboration for models of the pilot's neuromuscular subsystem in tracking tasks, Technical report, NASA.

Mainprice, J., Hayne, R., and Berenson, D. (2015). Predicting human reaching motion in collaborative tasks using inverse optimal control and iterative replanning, in *IEEE International Conference on Robotics and Automation (ICRA)* (IEEE), pp. 885–892.

Mangun, G. and Hillyard, S. (1991). Modulations of sensory-evoked brain potentials indicate changes in perceptual processing during visual-spatial priming, *Journal of Experimental Psychology: Human Perception and Performance* **17**(4), 1057–1069.

Manjunatha, H., Jujjavarapu, S. S., and Esfahani, E. T. (2022). Transfer learning of motor difficulty classification in physical human-robot interaction using electromyography, *Journal of Computing and Information Science in Engineering,* **22**(5), 1–32.

Markram, H. (2012). The human brain project, *Scientific American* **306**(6), 50–55.

Martínez-García, M., Gordon, T., and Shu, L. (2017). Extended crossover model for human-control of fractional order plants, *IEEE Access* **5**, 27622–27635.

Masarati, P., Quaranta, G., Bernardini, A., and Guglieri, G. (2015). Voluntary pilot action through biodynamics for helicopter flight dynamics simulation, *Journal of Guidance, Control, and Dynamics* **38**(3), 431–441.

Masarati, P., Quaranta, G., Zaichik, L., Yashin, Y., Desyatnik, P., Pavel, M. D., et al. (2013). Biodynamic pilot modelling for aeroelastic a/rpc, in *European Rotorcraft Forum* (Moscow, Russia).

Matsumoto, Y. and Griffin, M. J. (1998). Movement of the upper-body of seated subjects exposed to vertical whole-body vibration at the principal resonance frequency, *Journal of Sound and Vibration* **215**(4), pp. 743–762.

Matsumoto, Y. and Griffin, M. J. (2001). Modelling the dynamic mechanisms associated with the principal resonance of the seated human body, *Clinical Biomechanics* **16**(1), S31–S44.

Matsuo, Y., LeCun, Y., Sahani, M., Precup, D., Silver, D., Sugiyama, M., Uchibe, E., and Morimoto, J. (2022). Deep learning, reinforcement learning, and world models, *Neural Networks* **152**, 267–275.

McRuer, D. T. (1965). Human pilot dynamics in compensatory systems, Technical report, System technology Inc. Hawthorne Ca.

McRuer, D. T. and Jex, H. R. (1967). A review of quasi-linear pilot models, *IEEE Transactions on Human Factors in Electronics*, 3, 231–249.

McRuer, D. T. and Krendel, E. S. (1974). Mathematical models of human pilot behavior, AGARDograph AGARD-AG-188, pp. 1–83.

McRuer, D. T., Magdaleno, R. E., and Moore, G. P. (1968). A neuromuscular actuation system model, *IEEE Transactions on Man-Machine Systems* **9**(3), 61–71.

Menner, M., Worsnop, P., and Zeilinger, M. N. (2019). Constrained inverse optimal control with application to a human manipulation task, in *IEEE Transactions on Control Systems Technology* (IEEE).

Miall, R. C., Weir, D. J., Wolpert, D. M., and Stein, J. F. (1993). Is the cerebellum a smith predictor? *Journal of motor behavior* **25**(3), 203–216.

Milner, T. E. and Cloutier, C. (1993). Compensation for mechanically unstable loading in voluntary wrist movement, *Experimental Brain Research* **94**, 522–532.

Miranda, G. H. B. and Felipe, J. C. (2015). Computer-aided diagnosis system based on fuzzy logic for breast cancer categorization, *Computers in Biology and Medicine* **64**, 334–346.

Mishra, N. and Vaz, A. (2017). Bond graph modeling of a 3-joint string-tube actuated finger prosthesis, *Mechanism and Machine Theory* **117**, 1–20.

Miyoshi, T., Tano, S., Kato, Y., and Arnould, T. (1993). Operator tuning in fuzzy production rules using neural networks, in *[Proceedings 1993] Second IEEE International Conference on Fuzzy Systems*, pp. 641–646.

Mnih, V., Kavukcuoglu, K., Silver, D., Rusu, A. A., Veness, J., Bellemare, M. G., Graves, A., Riedmiller, M., Fidjeland, A. K., Ostrovski, G., *et al.* (2015). Human-level control through deep reinforcement learning, *Nature* **518**(7540), 529–533.

Mokhtari, A., M'Sirdi, N. K., Meghriche, K., and Belaidi, A. (2006). Feedback linearization and linear observer for a quadrotor unmanned aerial vehicle, *Advanced Robotics* **20**(1), 71–91.

Morasso, P. G. and Schieppati, M. (1999). Can muscle stiffness alone stabilize upright standing, *Journal of Neurophysiology* **82**, 1622–1626.

Mori, R. and Suzuki, S. (2009). Neural network modeling of lateral pilot landing control, *Journal of Aircraft* **46**(5), 1721–1726.

Morris, B. T. and Trivedi, M. M. (2011). Trajectory learning for activity understanding: Unsupervised, multilevel, and long-term adaptive approach, *IEEE Transactions on Pattern Analysis and Machine Intelligence* **33**(11), 2287–2301.

Mu, C., Wang, K., and Qiu, T. (2020). Dynamic event-triggering neural learning control for partially unknown nonlinear systems, *IEEE Transactions on Cybernetics* **52**(4), 2200–2213.

Mulder, M., Kaljouw, W. J., and Van Paassen, M. M. (2005). Parameterized multi-loop model of pilot's use of central and peripheral visual motion cues, in *AIAA Modeling and Simulation Technologies Conference and Exhibit*, AIAA-2005-5894.

Murphy, B. P. and Alambeigi, F. (2023). A surgical robotic framework for safe and autonomous data-driven learning and manipulation of an unknown deformable tissue with an integrated critical space, *Journal of Medical Robotics Research*, 2340001–2340016.

Murray-Smith, R. (1994). A local model network approach to nonlinear modelling, *University of Strathclyde*. DOI: 10.48730/me80-tw49.

Na, S.-H. and Lee, G.-E. (2021). Fuzzy fmea for rotorcraft landing system, *Journal of the Korea Academia-Industrial Cooperation Society* **22**(1), 751–758.

Nelles, O. (1996). Local linear model trees for on-line identification of time-variant nonlinear dynamic systems, in *International Conference on Artificial Neural Networks*, pp. 115–120.

Nepomuceno, E. and Martins, S. (2016). A lower bound error for free-run simulation of the polynomial narmax, *Systems Science & Control Engineering* **4**(1), 50–58.

Neto, P., Simão, M., Mendes, N., and Safeea, M. (2019). Gesture–based human–robot interaction for human assistance in manufacturing, *The International Journal of Advanced Manufacturing Technology* **101**, 119–135.

Ngo, H. Q. T., Le, V. N., Thien, V. D. N., Nguyen, T. P., and Nguyen, H. (2020). Develop the socially human–aware navigation system using dynamic window approach and optimize cost function for autonomous medical robot, *Advances in Mechanical Engineering* **12**(12), 1687814020979430.

Ngo, H. Q. T., Tran, A. S., Dong, V. K., and Yan, J. (2022). Implementation of the mathematical model for service robot to avoid obstacles and human, in *Proceedings of the Future Technologies Conference (FTC) 2021*, Vol. 2 (Springer), pp. 513–525.

Nguyen, A. H., Mai, L., and Do, H. N. (2020). Visual object tracking method of spatio-temporal context learning with scale variation, in *International Conference on the Development of Biomedical Engineering in Vietnam* (Springer, Cham), pp. 733–742.

Nguyen, K. and Kim, S. (2023). Peak-to-peak stabilization of sampled-data systems subject to actuator saturation and its practical application to an inverted pendulum, *Mathematics* **11**(22), 4592.

Nguyen, T. and Luo, X. (2021). Hybrid control systems for human-robot collaboration: A survey, *Robotics and Computer-Integrated Manufacturing* **67**, 101991.

Nicola, G. and Ghidoni, S. (2021). Deep reinforcement learning for motion planning in human robot cooperative scenarios, in *Proceedings of the 26th IEEE International Conference on Emerging Technologies and Factory Automation (ETFA)*.

Nicola, G., Villagrossi, E., and Pedrocchi, N. (2022). Human-robot co-manipulation of soft materials: Enable a robot manual guidance using a depth map feedback, in *Proceedings of the 31st IEEE International Conference on Robot and Human Interactive Communication (RO-MAN)*, pp. 498–504.

Nilsson, J., Silvlin, J., Brannstrom, M., Coelingh, E., and Fredriksson, J. (2016). If, when, and how to perform lane change maneuvers on highways, *IEEE Intelligent Transportation Systems Magazine* **8**(4), 68–78.

Obo, T., Loo, C. K., and Kubota, N. (2015). Robot posture generation based on genetic algorithm for imitation, in *2015 IEEE Congress on Evolutionary Computation (CEC)* (IEEE), pp. 552–557.

Oli, S., L'Esperance, B., and Gupta, K. (2013). Human motion behaviour aware planner (hmbap) for path planning in dynamic human environments, in *2013 16th International Conference on Advanced Robotics (ICAR)* (IEEE), pp. 1–7.

Oyelade, O. N., Ezugwu, A. E., Almutairi, M. S., Saha, A. K., Abualigah, L., and Chiroma, H. (2022). A generative adversarial network for synthetization of regions of interest based on digital mammograms, *Scientific Reports* **12**(1), 1–30.

Padfield, G. D. (1998). The making of helicopter flying qualities: A requirements perspective, *The Aeronautical Journal* **102**(1018), 409–437.

Pang, M. and Guo, S. (2013). A novel method for elbow joint continuous prediction using emg and musculoskeletal model, in *Proceedings of IEEE International Conference on Robotics and Biomimetics (ROBIO)*, pp. 1240–1245.

Pang, M., Guo, S., Huang, Q., Ishihara, H., and Hirata, H. (2015). Electromyography-based quantitative representation method for upper-limb elbow joint angle in sagittal plane, *Journal of Medical and Biological Engineering* **35**, 165–177.

Pattipati, K. R., Kleinman, D. L., and Ephrath, A. R. (1983). A dynamic decision model of human task selection performance, *IEEE Transactions on Systems, Man, and Cybernetics* **13**(2), 145–166.

Pavel, M. D., Jump, M., Dang-Vu, B., Masarati, P., Gennaretti, M., Ionita, A., *et al.* (2013). Adverse rotorcraft pilot couplings - past, present and future challenges, *Progress in Aerospace Sciences* **62**, 1–51.

Pavel, M. D., Masarati, P., Gennaretti, M., Jump, M., Zaichik, L., Dang-Vu, B., Lu, L., Yilmaz, D., Quaranta, G., Ionita, A., *et al.* (2015). Practices to identify and preclude adverse aircraft-and-rotorcraft-pilot couplings–a design perspective, *Progress in Aerospace Sciences* **76**, 55–89.

Payne, P. and Stech, E. (1969). Dynamic models of the human body technical report, sep. 1962-feb. 1964, Technical report, AMRL.

Peng, J., Liao, Z., Su, Z., Yao, H., Zeng, Y., and Dai, H. (2024). A dual closed-loop control strategy for human-following robots respecting social space, in *Proceedings - IEEE International Conference on Robotics and Automation*, pp. 11252–11258.

Phatak, A., Weinert, H., Segall, I., and Day, C. N. (1976). Identification of a modified optimal control model for the human operator, *Automatica* **12**(1), 31–41.

Piccardi, C. (2001). Controlling chaotic oscillations in delay-differential systems via peak-to-peak maps, *IEEE Transactions on Circuits and Systems I: Fundamental Theory and Applications* **48**(8), pp. 1032–1037.

Piccardi, C. (2008). Parameter estimation for systems with peak-to-peak dynamics, *International Journal of Bifurcation and Chaos* **18**(03), 745–753.

Piccardi, C. and Feichtinger, G. (2002). Peak-to-peak dynamics in the dynastic cycle, *Chaos, Solitons & Fractals* **13**(2), 195–202.

Pizzolato, C., Lloyd, D. G., Barrett, R. S., Cook, J. L., Zheng, M. H., Besier, T. F., and Saxby, D. J. (2017). Bioinspired technologies to connect musculoskeletal mechanobiology to the person for training and rehabilitation, *Frontiers in Computational Neuroscience* **11**, 96.

Pizzolato, C., Saxby, D. J., Palipana, D., Diamond, L. E., Barrett, R. S., Teng, Y. D., and Lloyd, D. G. (2019). Neuromusculoskeletal modeling-based prostheses for recovery after spinal cord injury, *Frontiers in Neurorobotics* **13**, 97.

Previc, F. H. and Ercoline, W. R. E. (2004). *Spatial Disorientation in Aviation* (Vol. 203) (AIAA).

Prinzmetal, W., McCool, C., and Park, S. (2005). Attention: Reaction time and accuracy reveal different mechanisms, *Journal of Experimental Psychology: General* **134**(1), 73–92.

Purves, D. and Lotto, R. B. (2011). *Why We See What We Do Redux: A Wholly Empirical Theory of Vision* (Sinauer Associates, Sunderland).

Raffin, A. and Stulp, F. (2019). Deep reinforcement learning with model predictive control for continuous human-robot interaction, in *Proceedings of IEEE/RSJ International Conference on Intelligent Robots and Systems (IROS)*, pp. 1243–1248.

Ragazzini, J. R. (1948). Engineering aspects of the human being as a servomechanism, in *Proceedings of the Meeting of the American Psychological Association*.

Ramadan, A., Choi, J., Radcliffe, C. J., Popovich, J. M., and Reeves, N. P. (2019). Inferring control intent during seated balance using inverse model predictive control, *IEEE Robotics and Automation Letters* **4**(2), 224–230.

Raney, D. L., Jackson, E. B., and Buttrill, C. S. (2002). Simulation study of impact of aeroelastic characteristics on flying qualities of a high-speed civil transport. Technical report, NASA.

Raney, D. L., Jackson, E. B., Buttrill, C. S., and Adams, W. M. (2001). The impact of structural vibration on flying qualities of a supersonic transport, in *AIAA Atmospheric Flight Mechanics Conference* (AIAA).

Rashidinejad, A., Nikravesh, S., and Talebi, H. (2015). Nonlinear bilateral teleoperation with flexible-link slave manipulator, in *Proceedings of the 3rd RSI International Conference on Robotics and Mechatronics (ICROM)*, pp. 284–289.

Rasmussen, J. (1983). Skills, rules, and knowledge; signals, signs, and symbols, and other distinctions in human performance models, *IEEE Transactions on Systems, Man, and Cybernetics* **3**, 257–266.

Ray, M., Mahata, N., and Sing, J. K. (2023). Uncertainty parameter weighted entropy-based fuzzy c-means algorithm using complemented membership functions for noisy volumetric brain mr image segmentation, *Biomedical Signal Processing and Control* **85**, 104925–104937.

Roth, W., Dorato, K., and Kopell, B. (1984). Intensity and task effects on evoked physiological responses to noise bursts, *Psychophysiology* **21**(3), 325–331.

Roy, A., Hossain, M., and Muromachi, Y. (2022). A deep reinforcement learning-based intelligent intervention framework for real-time proactive road safety management, *Accident Analysis and Prevention* **165**, 106512.

Rusanovsky, M., Beeri, O., and Oren, G. (2022). An end-to-end computer vision methodology for quantitative metallography, *Scientific Reports* **12**(1), 1–27.

Sadrfaridpour, B., Saeidi, H., Burke, J., Madathil, K., and Wang, Y. (2016). Modeling and control of trust in human-robot collaborative manufacturing, in *Robust Intelligence and Trust in Autonomous Systems* (Springer), pp. 115–141.

Safavi, A. and Zadeh, M. H. (2017). Teaching the user by learning from the user: Personalizing movement control in physical human-robot interaction, *IEEE/CAA Journal of Automatica Sinica* **4**(4), 704–713.

Salmanpour, M. R., Shamsaei, M., Saberi, A., Klyuzhin, I. S., Tang, J., Sossi, V., and Rahmim, A. (2020). Machine learning methods for optimal prediction of motor outcome in parkinson's disease, *Physica Medica* **69**, 233–240.

Salvendy, G. (1987). *Handbook of Human Factors* (John Wiley & Sons, New York).

Sanger, T. D. (1994). Neural network learning control of robot manipulators using gradually increasing task difficulty, *IEEE Transactions on Robotics and Automation* **10**, 323–333.

Schäfer, A. M. and Zimmermann, H.-G. (2007). Recurrent neural networks are universal approximators, *International Journal of Neural Systems* **17**(04), 253–263.

Schmidt, D. K. (1985). Pilot modeling and closed-loop analysis of flexible aircraft in the pitch tracking task, *Journal of Guidance, Control, and Dynamics* **8**(1), 56–61.

Schmidt, D. K. and Bacon, B. J. (1983). An optimal control approach to pilot/vehicle analysis and the neal-smith criteria, *Journal of Guidance, Control, and Dynamics* **6**(5), 339–347.

Schwartz, D., Guleyupoglu, B., Koya, B., Stitzel, J. D., and Gayzik, F. S. (2015). Development of a computationally efficient full human body finite element model, *Traffic Injury Prevention* **16**(sup1), S49–S56.

Schönfeld, A. (2010). Modified optimal control model and wake vortex encounter, in *Presentation at the WakeNet-Europe Specific Workshop: Models and Methods for WVE Simulations* **3**.

Scibilia, A., Laghi, M., De Momi, E., Peternel, L., and Ajoudani, A. (2018). A self-adaptive robot control framework for improved tracking and interaction performances in low-stiffness teleoperation, in *Proceedings of the 18th International Conference on Humanoid Robots (Humanoids)*, pp. 280–283.

Scibilia, A., Pedrocchi, N., and Fortuna, L. (2022a). Human control model estimation in physical human–machine interaction: A survey, *Sensors* **22**(5), 1732–1758.

Scibilia, A., Pedrocchi, N., and Fortuna, L. (2022b). Modeling of control delay in human-robot collaboration, in *Proceedings of the 48th Annual Conference of the IEEE Industrial Electronics Society*.

Scibilia, A., Pedrocchi, N., and Fortuna, L. (2024a). A nonlinear modeling framework for force estimation in human-robot interaction, *IEEE Access* **12**, 45–60.

Scibilia, A., Prini, A., Dinon, T., Pedrocchi, N., and Caimmi, M. (2024b). Over three decades of upper-limb robotic neurorehabilitation: Drawing conclusions and future work, in *2024 IEEE 20th International Conference on Automation Science and Engineering (CASE)*, pp. 1303–1310.

Self, R., Abudia, M., Mahmud, S. N., and Kamalapurkar, R. (2022). Model-based inverse reinforcement learning for deterministic systems, *Automatica* **140**, 110242–110255.

Shen, G., Jiao, Y., Lin, Y., and Huang, J. (2022). Approximation with cnns in sobolev space: with applications to classification, *Advances in Neural Information Processing Systems* **35**, 2876–2888.

Shi, W., Huang, Z., Huang, H., Hu, C., Chen, M., Yang, S., and Chen, H. (2022). Loen: Lensless optoelectronic neural network empowered machine vision, *Light: Science & Applications* **11**(1), 1–12.

Sirouspour, M. R. and Salcudean, S. E. (2003). Suppressing operator-induced oscillations in manual control systems with movable bases, *IEEE Transactions on Control Systems Technology* **11**(4), 448–459.

Sirouspour, S. (2005). Modeling and control of cooperative teleoperation systems, *IEEE Transactions on Robotics* **21**(6), 1220–1225.

Sirouspour, S. and Setoodeh, P. (2005). Multi-operator/multi-robot teleoperation: an adaptive nonlinear control approach, in *Proceedings of the IEEE/RSJ International Conference on Intelligent Robots and Systems*, pp. 1576–1581.

Sitole, S. P. and Sup, F. C. (2023). Continuous prediction of human joint mechanics using emg signals: A review of model-based and model-free approaches, *IEEE Transactions on Medical Robotics and Bionics* **5**(3), 528–546.

Smith, R. H. (1975). A theory for handling qualities with applications to mil-f-8785b, Technical report AFFDL-TR-75-119, WPAFB, OH: Air Force Flight Dynamics Lab.

Stewart, P., Gladwin, D., Parr, M., and Stewart, J. (2010). Multi-objective evolutionary fuzzy augmented flight control for an f16 aircraft, *Proceedings of the Institution of Mechanical Engineers, Part G: Journal of Aerospace Engineering* **224**(3), 293–309.

Strogatz, S. H. (1994). Norbert wiener's brain waves, in *Frontiers in Mathematical Biology* (Springer, Berlin), pp. 122–138.

Su, X., Wu, Y., Song, J., and Yuan, P. (2018). A fuzzy path selection strategy for aircraft landing on a carrier, *Applied Sciences* **8**(5), 779–796.

Sutton, R. S. and Barto, A. G. (2018). *Reinforcement Learning: An Introduction* (MIT Press).

Suzuki, S. and Furuta, K. (2012). Adaptive impedance control to enhance human skill on a haptic interface system, *Journal of Control Science and Engineering* **2012**.

Szczurek, K. A., Prades, R. M., Matheson, E., Rodriguez-Nogueira, J., and Di Castro, M. (2022). Mixed reality human-robot interface with adaptive communications congestion control for the teleoperation of mobile redundant manipulators in hazardous environments, *IEEE Access* **10**, 87182–87216.

Sövényi, S. and Gillespie, R. B. (2007). Cancellation of biodynamic feedthrough in vehicle control tasks, *IEEE Transactions on Control Systems Technology* **15**(6), 1018–1029.

Takacs, A., Kovacs, L., Rudas, I., Precup, R. E., and Haidegger, T. (2005). Models for force control in telesurgical robot systems, *Acta PolytechnicaHungarica* **12**(8), 95–114.

Takens, F. (1981). *Dynamical Systems and Turbulence* (Warwick, 1980), pp. 366–381.

Tan, W., Qu, X., and Wang, W. (2003). Compared with human pilot model of neural networks and quasi-linearity in frequency domain, *Acta Aeronautica et Astronautica Sinica* **24**(6), 4–12.

Telban, R., Cardullo, F., and Guo, L. (2000). Investigation of mathematical models of otolith organs for human centered motion cueing algorithms, in *Modeling and Simulation Technologies Conference*, Vol. 4291.

Todorov, E. and Li, W. (2005). A generalized iterative lqg method for locally-optimal feedback control of constrained nonlinear stochastic systems, in *Proceedings of the 2005, American Control Conference*, pp. 300–306.

Toma, M., Njilie, F., Ghajari, M., and Galvanetto, U. (2010). Assessing motorcycle crash-related head injuries using finite element simulations, *International Journal of Simulation Modelling* **9**(3), 143–151.

Tsang, P. S. and Vidulich, M. (2002). *Principles and Practice of Aviation Psychology* (CRC Press, Boca Raton).

Tustin, A. (1947). The nature of the operator's response in manual control, and its implications for controller design, *Journal of the Institution of Electrical Engineers–Part IIA: Automatic Regulators and Servo Mechanisms* **94**(2), 190–206.

Ulbrich, S. and Maurer, M. (2015). Towards tactical lane change behavior planning for automated vehicles, in *Proceedings of the 18th International Conference on Intelligent Transportation Systems*, pp. 989–995.

Ullrich, D., Butz, A., and Diefenbach, S. (2021). The development of overtrust: An empirical simulation and psychological analysis in the context of human-robot interaction, *Frontiers in Robotics and AI* **8**, p. 554578.

Uppal, S. K. and Vaz, A. (2022). Motion control of a phalange using tendon-based actuation system: A bond graph approach (Springer, Singapore), pp. 1479–1486.

Valero-Cuevas, F. J., Hoffmann, H., Kurse, M. U., Kutch, J. J., and Theodorou, E. A. (2009). Computational models for neuromuscular function, *IEEE Reviews in Biomedical Engineering* **2**, 110–135.

Van Paasen, M. M., Vaart, V. D., C., J., and Mulder, J. A. (2004). Model of the neuromuscular dynamics of the human pilot's arm, *Journal of Aircraft* **41**(6), 1482–1490.

Vasconez, J. P., Carvajal, D., and Cheein, F. A. (2019). On the design of a human-robot interaction strategy for commercial vehicle driving based on human cognitive parameters, *Advances in Mechanical Engineering* **11**(7), 1687814019862715.

Vaz, A., Singh, K., and Dauphin-Tanguy, G. (2015). Bond graph model of extensor mechanism of finger based on hook-string mechanism, *Mechanism and Machine Theory* **91**, 187–208.

Venrooij, J., Mulder, M., van Paassen, M. M., Mulder, M., and Abbink, D. A. (2009). Relating biodynamic feedthrough to neuromuscular admittance, in *SMC*, pp. 1668–1673.

Venrooij, J., Van Paassen, M. M., Mulder, M., Abbink, D. A., Mulder, M., Helm, V. D., Frans, C., *et al.* (2014a). A framework for biodynamic feedthrough analysis-part I: Theoretical foundations, *IEEE Transactions on Cybernetics* **44**(9), 1686–1698.

Venrooij, J., Van Paassen, M. M., Mulder, M., Abbink, D. A., Mulder, M., Helm, V. D., Frans, C., *et al.* (2014b). A framework for biodynamic feedthrough analysis-part ii: Validation and application, *IEEE Transactions on Cybernetics* **44**(9), 1699–1710.

Vezin, P. and Verriest, J. P. (2005). Development of a set of numerical human models for safety, in *Proceedings of the 19th International Technical Conference on the Enhanced Safety of Vehicles (ESV)*, 05–0163, pp. 16–32.

Viharos, Z. J. and Kis, K. B. (2015). Survey on neuro-fuzzy systems and their applications in technical diagnostics and measurement, *Measurement* **67**, 126–136.

von Merten, K. (2006). Using humos2 model for the reconstruction of accidents with thoracical injuries, *Journal of Biomechanics* **39**(Supplement 1), S165–S175.

Wang, C. *et al.* (2008). A revised optimal control pilot model for computer simulation, in *2nd International Conference on Bio-Informatics and Biomedical Engineering* (IEEE).

Wang, H., Liu, P. X., and Liu, S. (2017). Adaptive neural synchronization control for bilateral teleoperation systems with time delay and backlash-like hysteresis, *IEEE Transactions on Cybernetics* **47**(10), 3018–3026.

Wang, J., Pradhan, M. R., and Gunasekaran, N. (2022a). Machine learning–based human-robot interaction in its, *Information Processing & Management* **59**(1), 102750.

Wang, K., Su, J., Zhang, P., Huang, B., and Feng, W. (2021a). Interaction design of display and control equipment based on man–machine–environment system engineering, in *International Conference on Man–Machine–Environment System Engineering* (Springer), pp. 869–877.

Wang, R., Zhu, K., Zhang, Z., Li, J., and Pan, L. (2021b). Extracting control information from multi-channel surface emg signals to drive a musculoskeletal model, in *Proceedings of 27th International Conference on Mechatronics and Machine Vision in Practice*, pp. 253–257. DOI: 10.1109/M2VIP49856.2021.9665114.

Wang, Y., Wang, L., Guo, J., Papamichail, I., Papageorgiou, M., Wang, F.-Y., Bertini, R., Hua, W., and Yang, Q. (2022b). Ego-efficient lane changes of connected and automated vehicles with impacts on traffic flow, *Transportation Research Part C: Emerging Technologies* **138**, 103478–103503.

Wei, L. and Zhang, W. (2018). Collaborative human-robot assembly with model predictive control, *IEEE Transactions on Robotics* **34**(6), 1124–1137.

Wei, L.-Y., Chen, T.-L., and Ho, T.-H. (2011). A hybrid model based on adaptive-network-based fuzzy inference system to forecast taiwan stock market, *Expert Systems with Applications* **38**(11), 13625–13631.

Wei, Q. and Lu, H. (2018). A survey on model predictive control for human-robot interaction, *Control Engineering Practice* **72**, 53–68.

Wei, Z., Zhang, Z.-Q., and Xie, S. Q. (2024). Continuous motion intention prediction using semg for upper-limb rehabilitation: A systematic review of model-based and model-free approaches, *IEEE Transactions on Neural Systems and Rehabilitation Engineering.* **32**, pp. 1487–1504.

Wen, S., Wang, T., and Tao, S. (2022). Hybrid cnn-lstm architecture for lidar point clouds semantic segmentation, *IEEE Robotics and Automation Letters* **7**(3), 5811–5818.

Wierenga, R. D. (1969). An evaluation of a pilot model based on kalman filtering and optimal control, *IEEE Transactions on Man-Machine Systems* **10**(4), 108–117.

Wohler, M., Loy, F., and Schulte, A. (2014). Mental models as common ground for human-agent interaction in cognitive assistant systems, in *Proceedings of the International Conference on Human-Computer Interaction in Aerospace*, Article No. 20. DOI: 10.1145/2669592.2669686.

Wolpert, D. M., Ghahramani, Z., and Jordan, M. I. (1995). An internal model for sensorimotor integration, *Science* **269**(5232), 1880–1882.

Wu, C. M., Schulz, E., Pleskac, T. J., and Speekenbrink, M. (2022). Time pressure changes how people explore and respond to uncertainty, *Scientific Reports* **12**(1), 1–14.

Wu, G., Wu, Y., Lu, X., Xu, S., and Wang, C. (2020). Human–machine interface optimization design based on ecological interface design (EID) theory, in *Man–Machine–Environment System Engineering: Proceedings of the 20th International Conference on MMESE* (Springer), pp. 715–723.

Xian, B., Wang, S., and Yang, S. (2019). Nonlinear adaptive control for an unmanned aerial payload transportation system: Theory and experimental validation, *Nonlinear Dynamics* **98**(3), 1745–1760.

Xiao, Y., Liu, W., and Li, S. (2024). Model prediction based adaptive variable impedance control of manipulator with passivity, in *Chinese Control Conference (CCC)*, pp. 4535–4540.

Xiong, D., Zhang, D., Zhao, X., and Zhao, Y. (2021). Deep learning for emg-based human-machine interaction: A review, *IEEE/CAA Journal of Automatica Sinica* **8**(3), 512–533.

Xu, D., Wu, Q., and Zhu, Y. (2021). Development of a semg-based joint torque estimation strategy using hill-type muscle model and neural network, *Journal of Medical and Biological Engineering* **41**(1), 34–44.

Xu, K. (2012a). Visual scene simulation to civil aviation aircraft approaching and landing using dual fuzzy neural network, in *International Conference on Network Computing and Information Security*, pp. 275–280.

Xu, K., Zhang, G., and Xu, Y. (2012). A dual fuzzy neuro controller using genetic algorithm in civil aviation intelligent landing system, *Advanced Science Letters* **6**(1), 360–363.

Xu, K. J. (2012b). Decisional autonomy of approach and landing phase for civil aviation aircraft using dual fuzzy neural network, *Advanced Materials Research* **476**, 936–939.

Xu, L.-A., Liu, Z.-H., Li, X.-L., *et al.* (2008). Dynamic modeling and vibration characteristics of multi-dof upper part system of seated human body, *Chinese Journal of Engineering Design* **15**(4), 244–249.

Xu, S., Tan, W., Efremov, A. V., Sun, L., and Qu, X. (2017). Review of control models for human pilot behavior, *Annual Reviews in Control* **44**, 274–291.

Xu, S. and Wu, Y. (2021). Modeling multi-loop intelligent pilot control behavior for aircraft-pilot couplings analysis, *Aerospace Science and Technology* **112**, 106651–106663.

Yan, X., Kakadiaris, I. A., and Shah, S. K. (2014). Modeling local behavior for predicting social interactions towards human tracking, *Pattern Recognition* **47**(4), 1626–1641.

Yang, Y., Hua, C., and Guan, X. (2013). Adaptive fuzzy finite-time coordination control for networked nonlinear bilateral teleoperation system, *IEEE Transactions on Fuzzy Systems* **22**(3), 631–641.

Yang, Z., Guo, S., and Liu, Y. (2020). Comparison of isometric force estimation methods for upper limb elbow joints, in *2020 IEEE International Conference on Mechatronics and Automation (ICMA)*, pp. 1558–1563.

Yazdi, M., Daneshvar, S., and Setareh, H. (2017). An extension to fuzzy developed failure mode and effects analysis (fdfmea) application for aircraft landing system, *Safety Science* **98**, 113–123.

Ye, Y., Zhang, X., and Sun, J. (2019). Automated vehicle's behavior decision making using deep reinforcement learning and high-fidelity simulation environment, *Transportation Research Part C: Emerging Technologies* **107**, 155–170.

Yeo, K. and Melnyk, I. (2019). Deep learning algorithm for data-driven simulation of noisy dynamical system, *Journal of Computational Physics* **376**, 1212–1231.

Ying, L.-C. and Pan, M.-C. (2008). Using adaptive network based fuzzy inference system to forecast regional electricity loads, *Energy Conversion and Management* **49**(2), 205–211.

Zaal, P. M. T., Pool, D. M., Mulder, M., and Van Paassen, M. M. (2008). New types of target inputs for multi-modal pilot model identification, in *AIAA Modeling and Simulation Technologies Conference and Exhibit*, pp. 18–21.

Zanoni, A., Cocco, A., and Masarati, P. (2020). Multibody dynamics analysis of the human upper body for rotorcraft–pilot interaction, *Nonlinear Dynamics* **102**(3), 1517–1539.

Zaychik, K., Cardullo, F., and George, G. (2006). A conspectus on operator modeling: past, present and future, in *AIAA Modeling and Simulation Technologies Conference and Exhibit*, pp. 6625–6640.

Zeyada, Y. and Hess, R. A. (2003). Computer-aided assessment of flight simulator fidelity, *Journal of Aircraft* **40**(1), 173–180.

Zhang, H., Zhang, X., and Wang, J. (2014). Robust gain-scheduling energy-to-peak control of vehicle lateral dynamics stabilisation, *Vehicle System Dynamics* **52**(3), 345–360.

Zhang, J., Zhao, Y., Bao, T., Li, Z., Qian, K., Frangi, A. F., Xie, S. Q., and Zhang, Z.-Q. (2023a). Boosting personalized musculoskeletal modeling with physics-informed knowledge transfer, *IEEE Transactions on Instrumentation and Measurement* **72**.

Zhang, J., Zhao, Y., Bao, T., Li, Z., Qian, K., Frangi, A. F., Xie, S. Q., and Zhang, Z.-Q. (2023b). Physics-informed deep learning for musculoskeletal modeling: Predicting muscle forces and joint kinematics from surface emg,

IEEE Transactions on Neural Systems and Rehabilitation Engineering **31**, 484–493.

Zhang, J., Zhao, Y., Shone, F., Li, Z., Frangi, A. F., Xie, S. Q., and Zhang, Z.-Q. (2022a). Physics-informed deep learning for musculoskeletal modeling: Predicting muscle forces and joint kinematics from surface emg, *IEEE Transactions on Neural Systems and Rehabilitation Engineering* **31**, 484–493.

Zhang, S., Guo, S., Gao, B., Huang, Q., Pang, M., Hirata, H., and Ishihara, H. (2016). Muscle strength assessment system using semg-based force prediction method for wrist joint, *Journal of Medical and Biological Engineering* **36**, 121–131.

Zhang, T., Sun, H., and Zou, Y. (2022b). An electromyography signals-based human-robot collaboration system for human motion intention recognition and realization, *Robotics and Computer-Integrated Manufacturing* **77**, 102359–102374.

Zhang, Z., Cheng, S., and Solis-Lemus, C. (2022c). Towards a robust out-of-the-box neural network model for genomic data, *BMC Bioinformatics* **23**(1), 1–29.

Zhao, Y., Li, Q., Liu, Z., Alsaid, Y., Shi, P., and Jawed, M. K. (2023a). Sunlight-powered self-excited oscillators for sustainable autonomous soft robotics, *Science Robotics* **8**(67).

Zhao, Y., Li, Z., Zhang, Z., Qian, K., and Xie, S. (2023b). An emg-driven musculoskeletal model for estimation of wrist kinematics using mirrored bilateral movement, *Biomedical Signal Processing and Control* **81**.

Zhao, Y., Zhang, Z., Li, Z., Yang, A., Dehghani-Sanij, A., and Xie, S. (2020). An emg-driven musculoskeletal model for estimating continuous wrist motion, *IEEE Transactions on Neural Systems and Rehabilitation Engineering* **28**(12), 3113–3120.

Zhong, K., Yang, Z., Xiao, G., Li, X., Yang, W., and Li, K. (2021). An efficient parallel reinforcement learning approach to cross-layer defense mechanism in industrial control systems, *IEEE Transactions on Parallel and Distributed Systems* **33**(11), 2979–2990.

Zhu, F., Li, Y., Shi, Z., and Lin, H. (2022). Nonlinear identification and time-frequency domain analysis of corticomuscular responses during card-grabbing using the narmax method, in *Proceedings of the Control Automation Robotics Conference*.

Zhu, Q.-D., Li, H., Yu, M.-Z., Zhang, Z., and Jiang, X.-W. (2012). Landing risk evaluation of carrier-based aircraft based on bp neural network, in *2012 Second International Conference on Instrumentation, Measurement, Computer, Communication and Control*, pp. 1596–1601.

Zou, H., Tao, H., Zhou, Z., and Hu, B. (2022). Identification of mechanical impedance parameters of human upper limbs using mechanical perturbation method, in *Man–Machine–Environment System Engineering: Proceedings of the 21st International Conference on MMESE: Commemorative Conference for the 110th Anniversary of Xuesen Qian's Birth and the 40th Anniversary of Founding of Man–Machine–Environment System Engineering 21* (Springer), pp. 141–147.

Zube, A., Hofmann, J., and Frese, C. (2016). Model predictive contact control for human-robot interaction, in *47th International Symposium on Robotics, ISR 2016*, pp. 279–285.

9 789819 813612